Gas-Turbine Regenerators

Books of Related Interest

Centrifugal Pumps, Second Edition by I.J. Karassik and J.T. McGuire
Centrifugal Pump User's Guidebook by S. Yedidiah
Gears and Gear Manufacture by Richard H. Ewert
High Pressure Vessels by D.M. Fryer and J.F. Harvey
Theory and Design of Pressure Vessels by J.F. Harvey
Thermodynamics, Combustion and Engines by B.E. Milton
Foundations of Mechanical Engineering by K. Sherwin and A. Johnson
Strain Gauge Technology, Second Edition by A.L. Window
Principles of Turbomachinery, Second Edition by R.K. Turton

Gas-Turbine Regenerators

Douglas Stephen Beck
David Gordon Wilson

CHAPMAN & HALL

ITP® International Thomson Publishing
New York • Albany • Bonn • Boston • Cincinnati • Detroit • London • Madrid • Melbourne
Mexico City • Pacific Grove • Paris • San Francisco • Singapore • Tokyo • Toronto • Washington

Cover Design: Sa d Sayrafiezadeh, emDASH inc.
Art Direction: Andrea Meyer

Softcover reprint of the hardcover 1st edition 1996

For more information, contact:

Chapman & Hall
115 Fifth Avenue
New York, NY 10003

Chapman & Hall
2-6 Boundary Row
London SE1 8HN
England

Thomas Nelson Australia
102 Dodds Street
South Melbourne, 3205
Victoria, Australia

Chapman & Hall GmbH
Postfach 100 263
D-69442 Weinheim
Germany

International Thomson Editores
Campos Eliseos 385, Piso 7
Col. Polanco
11560 Mexico D.F.
Mexico

International Thomson Publishing-Japan
Hirakawacho-cho Kyowa Building, 3F
1-2-1 Hirakawacho-cho
Chiyoda-ku, 102 Tokyo
Japan

International Thomson Publishing Asia
221 Henderson Road #05-10
Henderson Building
Singapore 0315

1 2 3 4 5 6 7 8 9 10 XXX 01 00 99 97 96

Library of Congress Cataloging-in-Publication Data

Beck, Douglas Stephen
Gas-turbine regenerators / Douglas Stephen Beck, David Gordon Wilson
p. cm.
Includes bibliographical references and index.

ISBN-13: 978-1-4612-8512-0 e-ISBN-13: 978-1-4613-1209-3
DOI: 10.1007/978-1-4613-1209-3

1. Automotive gas turbines. 2. Gas-turbine power-plants. 3. Heat regenerators.
I. Wilson, David Gordon, 1928— . II. Title.
TL227.B43 1996
629.25 dc20 96-13362
CIP

British Library Cataloguing in Publication Data available

To order this or any other Chapman & Hall book, please contact **International Thomson Publishing, 7625 Empire Drive, Florence, KY 41042.** Phone: (606) 525-6600 or 1-800-842-3636.
Fax: (606) 525-7778, e-mail: order@chaphall.com.

For a complete listing of Chapman & Hall s titles, send your requests to
Chapman & Hall, Dept. BC, 115 Fifth Avenue, New York, NY 10003.

Contents

1 Introduction **1**

1.1 Regenerators vs. Recuperators 1
1.2 Heat Transfer 4
1.2.1 Regenerator Effectiveness 4
1.2.2 Heat-Capacity Rates 5
1.2.3 Convective Conductances and Number of Transfer Units . 7
1.2.4 Core Compactness 9
1.2.5 Core Rotation 10
1.2.6 Core Conduction 10
1.2.7 Porosity 14
1.2.8 Transient Operation 17
1.2.9 Flow Non-Uniformity 18
1.3 Leakage 22
1.4 Pressure Drops 24
1.5 Power Consumption 26
1.6 Summary 26

2 Background **27**

2.1 History of Regenerators 27
2.1.1 Early Regenerators 27
2.1.2 The Stirling Regenerator 28
2.1.3 The Siemens Regenerator 28
2.1.4 Cowper Stoves 28
2.1.5 The Ljungstrom Air-Preheater 30
2.1.6 Chrysler Turbine Cars 30
2.1.7 Chrysler-Corning Collaboration on Ceramics 32
2.1.8 Rover-Penny Turbine at Le Mans 32
2.1.9 Setback at Ford 33
2.2 Analysis and Design 33
2.2.1 Cooling Towers and the *H.T.U.* 33
2.2.2 Core Data 33
2.2.3 Design Optimization 34
2.2.4 State of the Art in 1953 34
2.2.5 Finite Core-Rotation Rate 34

2.2.6 Heat Conduction . . . 34
2.2.7 Flow Non-Uniformity . . . 35
2.3 State of the Art . . . 35

3 Gas-Turbine Cycles **37**
3.1 Performance . . . 37
3.2 Governing Equations . . . 40
3.3 Simple Cycle . . . 42
3.4 Regenerative Cycle . . . 52
3.5 Intercooled Regenerative (ICR) Cycle . . . 55
3.6 ICR Cycle with Reheat . . . 56
3.7 Summary . . . 62

4 Regenerator Designs **63**
4.1 Significance of Heat Exchangers in Gas Turbines . . . 63
4.1.1 Leakage . . . 64
4.1.2 Regenerator Advantages over Recuperators . . . 66
4.1.3 Size Limitations for Ceramic Rotary Regenerators . . . 67
4.2 Alternative Regenerator Designs . . . 67
4.2.1 Two-Chamber and Rotary Regenerators . . . 67
4.2.2 Alternative Rotary-Regenerator Configurations . . . 68
4.2.3 New Regenerator Concepts . . . 69
4.2.4 Discontinuous-Rotation Regenerator . . . 69
4.2.5 Modular Regenerator . . . 71
4.3 Gas-Turbine Cycles with Heat Exchangers . . . 72
4.3.1 The "Simple" Regenerative Cycle . . . 75
4.3.2 The Intercooled-Regenerative Cycle . . . 75
4.3.3 The Exhaust-Heated Cycle . . . 76
4.3.4 Solar-Heated Cycle . . . 76
4.4 Future Directions . . . 78

5 Design Procedures and Examples **79**
5.1 Direct Regenerator Design . . . 79
5.1.1 Specifications and Results . . . 80
5.1.2 Five-Step Procedure . . . 88
5.2 Optimal Regenerator Design . . . 101
5.3 Method of Kays and London . . . 105
5.3.1 Rotary Regenerator for a 1 MW Gas Turbine . . . 105
5.3.2 Modular Regenerator for a 10 MW Gas Turbine . . . 111
5.3.3 Introduction of Hot-Side Pressure Drop, Matrix Temperature Gradient, Cell Shape and Hydraulic Diameter as Design Parameters . . . 112
5.4 Summary . . . 118

6 Regenerator Performance **121**
6.1 Heat Transfer 121
6.1.1 Effectiveness Correlations 121
6.1.2 Physical Phenomena 124
6.1.3 Simplifying Assumptions 124
6.1.4 Governing Equations 126
6.1.5 Infinite Core-Rotation Rate 133
6.1.6 Finite Core-Rotation Rate 137
6.1.7 Axial Conduction during Flow Exposures 140
6.1.8 Heat Diffusion Under Seals 141
6.1.9 Seal-Width Effect 151
6.1.10 Porosity 151
6.1.11 Transient Operation 168
6.1.12 Seal Shape 180
6.1.13 Discontinuous Rotation 201
6.2 Summary 225
6.3 Seal Leakage 227
6.4 Pressure Drops 229

A Performance of the Ericsson Cycle **235**

Index **246**

Preface

Regenerative gas turbines are attractive alternatives to diesel engines and spark-ignition engines for automobiles and to diesel engines and combined-cycle engines for power generation. Theory indicates regenerative gas turbines should achieve higher thermal efficiencies than those of diesel engines and combined-cycle engines. Further, regenerative gas turbines are potentially lower in cost, require less maintenance, require less space, and pollute less than competitive systems.

Regenerators can be used for exhaust-gas heat exchange or for intercooling in gas-turbine systems. As an exhaust-gas heat exchanger, a regenerator recovers heat from the exhaust and uses it to preheat the compressed air before the compressed air enters the combustor. Preheating of the compressed air permits a small heat input to the combustor for a given power output of the engine. As an intercooler, a regenerator cools the gas between compressor stages. Less work is required to compress cool gas than is required to compress warm gas. Therefore, a regenerator intercooler can reduce the required work input to the compressor. Thus, regenerators can be used to increase the thermal efficiencies and power outputs of gas turbines.

High-performance regenerators are the backbones of high-performance regenerative gas turbines. In the past, lack of understanding of regenerator performance has led to sub-optimal engine designs. Now this book gives comprehensive regenerator information. With this book, the designer can design regenerators that will yield gas turbines with maximum thermal efficiencies.

This book contains both theoretical regenerator information and practical design examples. This book can be used by engineers in four ways. First, in preliminary stages of design, designers can choose from among the regenerator designs given in Chapter 4. The designer can identify one or more designs from among those given that may be suitable for the designer's application. Second, the designer who has some design specifications can use the optimization method given in Chapter 5 to choose the remaining specifications, such that a maximum thermal efficiency will be achieved for the resulting engine. For example, for fixed values of the compressor and turbine efficiencies, the turbine-inlet temperature (TIT), etc. a unique optimal regenerator design can be determined. Third, this book can be used by the manufacturer of regenerator cores. The regenerator-performance information given in Chapters 5 and 6 should be used to select core-passage geometries and core materials. Fourth, the inexperienced designer can use the step-by-step examples of designs of regenerative gas turbines given in Chapter 5.

This book is organized as follows. Some background is provided first. The introduction in Chapter 1 is intended to make the reader familar with the operation of regenerators and with the terminology that is used in regenerator analysis and design. Chapter 1 introduces all regenerator phenomena considered in this book, as well as relevant system parameters. Chapter 2 gives the historical background of regenerators. Historical background is given after the introduction so that the background will be more intelligible to the reader once the essential phenomena and terminology have been introduced.

Chapter 3 considers gas-turbine cycles, which are the thermodynamic cycles by which gas turbines convert heat into work. Regenerator performance consists of heat transfer, leakage, pressure drops, and power consumption. Chapter 3 considers the effects of each of these performances on the performance of regenerative gas-turbine cycles.

Chapter 4 describes various regenerator designs that are suitable for gas-turbine applications. Each kind of regenerator that is described has different performance characteristics and different sets of advantages and disadvantages.

Chapter 5 describes how to design high-performance gas-turbine regenerators. Three design methods are described: (1) Direct Regenerator Design, (2) Optimal Regenerator Design, and (3) the design method of Kays and London [37]. The first method directly determines regenerator dimensions from input performance specifications. The second method finds optimal values for pressure ratios and regenerator dimensions that maximize cycle thermal efficiency. The third method is appropriate for preliminary design. The first two design methods have been included in the commercial computer software RGT-OPT™ [1]

Chapter 6 consider three kinds of regenerator performance: (1) heat transfer, (2) leakage, and (3) pressure drops. We quantify these performances for regenerators operating under a wide range of conditions.

The authors hope this book will lead to high-performance gas-turbine regenerators and high-efficiency gas-turbine systems.

Douglas Stephen Beck
David Gordon Wilson
May 25, 1996

---

[1] RGT-OPT is a trademark of Douglas Stephen Beck, the principal author, and is available from him.

Acknowledgments

This book evolved from the PhD research of the principal author (Douglas Stephen Beck), which was supervised by the second author (David Gordon Wilson) at MIT. The principal author thanks Prof. Wilson for his guidance with the PhD research and for his contributions to Chapters 2, 4, and 5. The NASA Lewis Research Center partially funded the research under the NASA Graduate Student Researchers' Program and also under a separate grant. Paul Kerwin and Tom Strom of NASA Lewis provided technical advice.

Prof. Alex Brown (head of the mechanical, materials, and civil engineering school at the Royal Military College of Science) supported the principal author's research during the summer of 1992. Profs. A. Douglas Carmichael and Peter Griffith of MIT were the other members of the principal author's PhD thesis committee. Both provided suggestions for research directions. Mark Franchett of General Motors and Daniel Lipp of Corning assisted in the procurement of ceramic regenerator-core samples that were used in transient-performance experiments.

Some of the work discussed in Chapters 2, 4, and 5 was carried out at MIT with partial support by grants from the US Department of Energy (DoE) and by a contract with Lincoln Laboratory, MIT. A legion of former thesis students from BSME to PhD have contributed valuably to the program. A paper published by the Institution of Mechanical Engineers (Wilson [80]) was used freely for the historical background in Chapter 2.

The authors greatly appreciate all of these contributors.

Nomenclature

Roman-Letter Symbols

a	Intermediate parameter used to quantify core conduction (–)
A	Area (m^2)
$\mathbf{A}$	Matrix used in finite-difference integration schemes (–)
A_f	Total face area (m^2)
A_{ff}	Free-flow area (m^2)
A_h	Heat-transfer area (m^2)
As	Solid area (m^2)
As'	Solid-area ratio (–), the ratio of the solid core area on the compressed-air side to the solid core area on the exhaust side
$\mathbf{b}$	Vector used in finite-difference integration schemes (–)
BS/A	Dimensionless core-rotation speed (–)
c	Specific heat capacity ($J/kg - K$)
C	Heat-capacity rate (W/K)
C_f	Friction coefficient (–)
C_N	Heat-capacity rate of the compressed air (W/K)
C_R	Heat-capacity rate of the core material (W/K)
C_{RAT}	Heat-capacity-rate ratio (–), the ratio of the heat-capacity rate of the compressed air to the heat-capacity rate of the exhaust
C_{ROT}	Dimensionless core rotation rate (–), the ratio of the heat-capacity rate of the core material to the heat-capacity rate of the compressed air
C_X	Heat-capacity rate of the exhaust (W/K)
CC	Core Compactness (–)
D	Diameter (m)
D_H	Hydraulic diameter (m)
f	Friction factor (–)
f	Fuel-to-air ratio (–)
Fo	Fourier number (–), the ratio of the product of thermal diffusivity and time to the square of a characteristic length
h	Convective heat-transfer coefficient ($W/m^2 - K$)
H	One-half the thickness of a passage wall (m)
h_{ef}	Effective convective heat-transfer coefficient ($W/m^2 - K$)

Roman-Letter Symbols (cont.)

(hA)	Convective conductance (W/K)
$(hA)'$	Convective-conductance ratio (–), the ratio of the convective conductance on the compressed-air side to the convective conductance on the exhaust side
HRB	Heat-Recovery Boiler
HRSG	Heat-Recovery Steam Generator
i	Specific enthalpy (J/kg)
k	Thermal conductivity $(W/m-K)$
L	Core thickness or flow length (m)
L_S	Seal length (m)
m	Mass (kg)
$\dot{m}$	Mass flow rate (kg/s)
$(\dot{m}_{Xlc})_+$	Carry-over leakage associated with rotation from the compressed-air side to the exhaust side (kg/s)
$(\dot{m}_{Xlc})_-$	Carry-over leakage associated with rotation from the exhaust side to the compressed-air side (kg/s)
(mc)	Heat capacity (J/K)
n	Number of throttlings (–)
N	A number (–)
N_{Mo}	Mondt number (–)
N_p	Porosity number (–)
NTU	Number of Transfer Units (–), a dimensionless heat-exchanger size
Nu	Nusselt number (–)
p	Core porosity (–), the ratio of voids volume to total volume
$\wp$	Wetted perimeter (m)
P	Pressure (Pa)
Pe	Péclét number (–), equal to the product of Reynolds number and Prandlt number $(Pe = RePr)$
Pr	Prandlt number (–), the ratio of kinematic viscosity (m^2/s) to thermal diffusivity (m^2/s)
Q'	Cycle specific heat-input rate (–)
$\dot{Q}$	Heat flow (W)
(q/A)	Heat flux (W/m^2)
r	Pressure ratio (–)
r	Radius (m)
R	Radius (m)
R	Gas constant $(J/kg-K)$
Re	Reynolds number (–), the ratio of inertial effects to viscous effects in a flow
s	Specific entropy $(J/kg-K)$
S	Core rotational speed (m/s)
SC	Seal Coverage (–), the fraction of the core face area covered by seals
t	Time (s)
T	Absolute temperature (K)

Roman-Letter Symbols (cont.)

T	Time period (s)
T'	Temperature ratio (–), the ratio of the turbine inlet temperature (TIT) to the compressor inlet temperature
TIT	Turbine Inlet Temperature (K)
u	Local velocity (m/s)
u	Specific internal energy (J/kg)
U	Velocity (m/s)
v	Specific volume (m^3/kg)
V	Volume (m^3)
RPR	Rotation-period-to-Pause-period Ratio (–)
V_R	Total core volume (m^3)
V'	Core volume ratio (–), the ratio of the core volume on the compressed-air side to the volume on the exhaust side
$\dot{V}$	Volumetric flow rate (m^3/s)
w	Width (m)
w_S	Seal width (m)
W'	Cycle specific power (–)
$\dot{W}$	Power (W)
x	Length (m)
$\mathbf{x}$	Vector used in finite-difference integration schemes (–)
y	Height (m)
Y	One-half the spacing of parallel plates (m)
Y_i	Height of the inlet header (m)
Y_o	Height of the outlet header (m)

Greek-Letter Symbols

α	Flow coefficient (–)
α_R	Thermal diffusivity of the core material (m^2/s)
α_{ROT}	Discontinuous-rotation angle (degrees)
β	Half-angle of an annular sector (radians)
γ	Flow coefficient (–)
δ	Differential (–)
δ	Seal clearance (m)
Δ	Difference
Δ	Seal-location angle (degrees)
Δ	Intermediate parameter used for quantifying the effects of low porosity on regenerator effectiveness (–)
Δ_{AC}	Axial-conduction effect (–)
Δ_{DR}	Discontinuous-rotation effect (–)
Δ_{ROT}	Core-rotation effect (–)
Δ_S	Seal-width effect (–)
Δ_W	Uniform-seal-width effect (–)

Greek-Letter Symbols (cont.)

ϵ	Effectiveness (–), the ratio of the actual amount of heat transferred by a heat exchanger to the theoretical maximum that could be transferred by a heat exchanger of infinite size
ζ	Dimensionless length (–)
η	Efficiency (–)
η_{TH}	Thermal efficiency (–)
Θ	Dimensionless temperature (–)
Θ	Seal-location angle used for quantifying the effects of seal shape on regenerator effectiveness (radians)
κ	Permeability (m^2)
κ	Intermediate parameter used to quantify core conduction (–)
λ	Core-conduction parameter (–)
μ	Viscosity ($kg/m - s$)
ξ	Dimensionless factor used for converting governing equations in the time domain to equations for periodic steady-state rotation (–)
ρ	Density (kg/m^3)
τ	Time period (s)
τ_{ROT}	Rotation period (s)
τ_w	Skin-friction shear stress (Pa)
$\tau^\star$	Dimensionless time period (–)
ϕ	Intermediate angle used for quantifying the effects of seal shape on regenerator effectiveness (radians)
ω	Angular rotational speed (rad/s)

Roman-Letter Subscripts

AC	Associated with Axial Conduction
atm	Of the atmosphere
c	Carry-over
c	Across the core
C	Carry-over
C	Of the Compressor
CF	CounterFlow
$COND$	Associated with heat conduction
$cont$	Continuous
$crit$	Critical
d	Direct
D	Direct
$disc$	Discontinuous
E	Of the Expander or turbine
f	Associated with the core face
ff	Free flow

Roman-Letter Subscripts (cont.)

h	Associated with heat transfer
h	Humps
H	Of the Heater or combustor
H	Hydraulic
H	Based on the thickness of a passage wall
i	Inlet or inner
I	Inlet
I	Of the Intercooler
L	Based on core thickness or flow length, L
L	Leakage
m	Mixed mean
m	Mean or average
m	Of the core matrix
MAX	Maximum
N	On the side of miNimum heat-capacity rate (usually the compressed-air side in gas-turbine regenerators)
o	Outlet or outer
O	Outlet
P	At constant pressure
$pass$	Of a passage
R	Of the core, or of the regenerator
REF	Reference
ROT	Of rotation
S	Associated with the seals
$sect$	Of an annular sector
ss	Steady-state
st	Static
TH	Thermal
TOT	Total
V	At constant volume
w	Of a passage wall
W	Associated with a uniform-width seal
X	On the side of maXimum heat-capacity rate (usually the exhaust side in gas-turbine regenerators), or of the regenerator in general

Roman-Letter Superscripts

T	Transpose

Greek-Letter Subscripts

Σ	Sum or total

Numerical Subscripts

0	At the fluid/solid interface
0	At the inlet to the gas turbine
0	Total property
00	Total-to-total
1	At the compressor inlet
1	First (for example $C1$ indicates the first compressor)
1	In the fluid domain
2	At the point in the compressor discharge *before* the turbine coolant has been extracted
2	Second (for example $C2$ indicates the second compressor)
4	At the point in the compressor discharge *after* the turbine coolant has been extracted
9	At the turbine inlet
10	At the turbine outlet
13	Outlet of the regenerator on the compressed-air side
19	Discharge of the first compressor
22	Outlet of the intercooler

Miscellaneous Indicators

$\star$	Dimensionless
$+$	Associated with positive rotation from the compressed-air side to the exhaust side
$-$	Associated with negative rotation from the exhaust side to the compressed-air side

List of Figures

1.1 How Heat Exchangers Improve the Performances of Gas Turbines. 2
1.2 Schematic of an Axial-Flow Rotary-Regenerator Core. 3
1.3 Schematic of a Counter-Flow Recuperator Core. 3
1.4 Temperatures through a Regenerator Core. 5
1.5 Effect of Heat-Capacity-Rate Ratio (C_{RAT}) on Regenerator Effectiveness. Reprinted from Beck [7] with permission from the ASME. 6
1.6 Schematic of an Isolated Regenerator Passage Tube (The ratio, D_H/L, is much larger than what is typical for gas turbines). Reprinted from Beck [5] with permission from the ASME. 7
1.7 Temperatures through a Regenerator Core with Equal Heat-Capacity Rates ($C_{RAT} = 1$) and (a) a Perfect Effectiveness ($\epsilon \to 1$) and No Axial Conduction and (b) Infinite Conduction. 11
1.8 Schematic of an Isolated Passage Tube in a Parallel-Plate Regenerator. 16
1.9 (a) An Ideal Header/Core System and (b) a Non-Ideal Header/Core System. 19
1.10 (a) Core Passages with Uniform Geometries and (b) Core Passages with Non-Uniform Geometries. 20
1.11 Regenerator Seals of Alternative Shapes: (a) Annular-Sector Shaped Seals; and (b) Uniform-Width Seals. 21
1.12 Regenerator Core that Experiences Discontinuous Rotation. . . . 21
1.13 Typical Plot of Regenerator Leakage. 23

2.1 Cowper (or Hot) Stove (adapted from Reese [59]). 29
2.2 Ljungstrom Air Preheater (adapted from Babcock and Wilcox [1] with permission). 31
2.3 NGTE Rotary Regenerator with "Smart" Seals. 32

3.1 Simple-Cycle Gas Turbine. 42
3.2 Design-Point Performance and Plot of Temperature vs. Specific Entropy for a Simple-Cycle Gas Turbine with Component Specifications Listed in Table 3.1 (data calculated by RGT-OPT). . . 44

3.3 Design-Point Performance and Plot of Temperature vs. Specific Entropy for a Low-Pressure-Ratio Simple-Cycle Gas Turbine with Component Specifications Listed in Table 3.2 (data calculated by RGT-OPT). 51

3.4 Regenerative Gas Turbine. 53

3.5 Design-Point Performance and Plot of Temperature vs. Specific Entropy for a Regenerative Gas Turbine with Component Specifications Listed in Table 3.3 (data calculated by RGT-OPT). . . 54

3.6 Intercooled Regenerative (ICR) Gas Turbine. 56

3.7 Design-Point Performance and Plot of Temperature vs. Specific Entropy for an ICR Gas Turbine with Component Specifications Listed in Table 3.4 (data calculated by RGT-OPT). 57

3.8 ICR Gas Turbine with Reheat. 59

3.9 Design-Point Performance and Plot of Temperature vs. Specific Entropy for an ICR Gas Turbine with Reheat with Component Specifications Listed in Table 3.5 (data calculated by RGT-OPT). 60

4.1 Design-Point Performance of Gas-Turbine Cycles. This material has been reproduced from the Proceedings Part A Issue A3 1993 Volume 207 p. 196 Figure 3 by David Gordon Wilson by permission of the Council of the Institution of Mechanical Engineers, London. 65

4.2 Effect of Leakage on Thermal Efficiency of a Regenerative Cycle (adapted from Wilson and Beck [81]). 66

4.3 Propagation of a Thermocline through the Core of a Switching Regenerator. This material has been reproduced from the Proceedings Part A Issue A3 1993 Volume 207 p. 197 Figure 4 by David Gordon Wilson by permission of the Council of the Institution of Mechanical Engineers, London. 68

4.4 Alternative Forms of Rotary Regenerators. 70

4.5 Discontinuously Moving Regenerator with Clampable Seals. This material has been reproduced from the Proceedings Part A Issue A3 1993 Volume 207 p. 200 Figure 8 by David Gordon Wilson by permission of the Council of the Institution of Mechanical Engineers, London. 71

4.6 Sequence from Rotary-Disk to Linear Modular, Showing Reduction in Seal Length. 73

4.7 One Form of Modular Ceramic Regenerator. This material has been reproduced from the Proceedings Part A Issue A3 1993 Volume 207 p. 200 Figure 10 by David Gordon Wilson by permission of the Council of the Institution of Mechanical Engineers, London. 74

4.8 The Modular Regenerator in the Supplementary-Fired Exhaust-Heated Cycle with Design-Point Performance. This material has been reproduced from the Proceedings Part A Issue A3 1993 Volume 207 pp. 205 and 206 Figures 5 and 6 by David Gordon Wilson by permission of the Council of the Institution of Mechanical Engineers, London. 77
4.9 Modular Regenerator Used in a Solar-Heated System. 78

5.1 Dimensions of Regenerators with Radial Flows. 81
5.2 Dimensions of Regenerators with Axial Flows. 82
5.3 Effect of Core Rotation Rate on Regenerator Effectiveness (data from Bahnke and Howard [2]). 85
5.4 Regenerator Temperatures. 107
5.5 Modular-Regenerator Configuration. 112
5.6 Specifications for Design Example. 115

6.1 Isolated Regenerator Passage Tube. 127
6.2 Effect of Dimensionless Core Size (NTU) on Effectiveness (ϵ). . . 134
6.3 Temperature Distributions through a Regenerator Core for Flows with Equal Heat-Capacity Rates ($C_{RAT} = 1$). 134
6.4 Temperature Distributions through a Regenerator Core ($NTU = 1$). 135
6.5 Effectiveness for Finite Core-Rotation Rates ($C_{RAT} = 1$ and $(hA)' = 1$). 138
6.6 Effectiveness for Finite Core-Rotation Rates ($C_{RAT} = 0.9$ and $(hA)' = 1$). 138
6.7 Effectiveness for Finite Core-Rotation Rates ($C_{RAT} = 0.95$ and $(hA)' = 1$). 139
6.8 Solutions for One-Dimensional Heat Diffusion. Reprinted from Beck [7] with permission from the ASME. 143
6.9 Plot of all Sample-Calculation Results Given in this Section. . . . 144
6.10 Plot of $(C_{ROT}/\lambda)Fo_L$ vs. Seal Coverage. 145
6.11 Effects of Seal Width on Effectiveness for $NTU = 20$; $C_{ROT} = 3$; $C_{RAT} = (hA)' = As' = 1$; and $\lambda = 0.01$, 0.08, and 0.32. Reprinted from Beck [7] with permission from the ASME. 146
6.12 Effects of Seal Width on Effectiveness for $NTU = 20$; $C_{RAT} = (hA)' = As' = 1$; $\lambda = 0.01$; and $C_{ROT} = 1$, 3, and 5. Reprinted from Beck [7] with permission from the ASME. 147
6.13 Effects of Seal Width on Effectiveness for $C_{ROT} = 3$; $C_{RAT} = (hA)' = As' = 1$; $\lambda = 0.32$; and $NTU = 1$, 10, and 20. Reprinted from Beck [7] with permission from the ASME. 147
6.14 Effects of Seal Width on Effectiveness for $NTU = 20$; $C_{ROT} = 3$; $(hA)' = As' = 1$; $\lambda = 0.01$; and $C_{RAT} = 0.90$, 0.95, and 1.0. Reprinted from Beck [7] with permission from the ASME. 148

6.15 Effects of Seal Width on Effectiveness for $NTU = 20$; $C_{ROT} = 3$; $C_{RAT} = 1.0$; $\lambda = 0.32$; and $(hA)' = As' = 0.25$ and 1.0. Reprinted from Beck [7] with permission from the ASME. 149
6.16 Schematic of a Modular Regenerator. Reprinted from Beck [7] with permission from the ASME. 150
6.17 Effects of Low Porosity on the Heat-Transfer Performance of Regenerators (lower scales are for ceramic cores [$k/k_R = 0.19$] and stainless-steel cores [$k/k_R = 0.0022$]). 152
6.18 Isolated Half-Passage-Tube of a Parallel-Plate Regenerator. . . . 153
6.19 Core-Material Temperature Profiles for $N_p = 0.1$ and $0 \leq Fo_H \leq 0.1$. 157
6.20 Core-Material Temperature Profiles for $N_p = 0.1$ and $0.1 \leq Fo_H \leq 1$. 157
6.21 Core-Material Temperature Profiles for $N_p = 0.01$ and $0 \leq Fo_H \leq 0.1$. 158
6.22 Core-Material Temperature Profiles for $N_p = 0.01$ and $0.1 \leq Fo_H \leq 1$. 159
6.23 Reduction in Core Heat-Transfer Performance vs. Dimensionless Time for $N_p = 0.1$. 159
6.24 Reduction in Core Heat-Transfer Performance vs. Dimensionless Time for $N_p = 0.01$. 160
6.25 Core-Material Temperature Profiles for $N_p = 1$ and $0.1 \leq Fo_H \leq 1$. 160
6.26 Normalized Reductions in Core Heat-Transfer Performance vs. Dimensionless Time for Various N_p. 161
6.27 C_X-Side Outlet-Temperature Response of a Regenerator with $NTU = 10$; $C_{ROT} = 0.35$; $C_{RAT} = 1.0$; $(hA)' = 1.0$; $As' = 1.0$; $\lambda = 0.01$; and $SC = 0.2$ and a Model Temperature Response. . . 162
6.28 Reductions in Core Heat-Transfer Performance for Various Flow-Temperature Dynamics. 163
6.29 Temperature Profiles for $Fo_H < 0.1$ in a Core Wall with $N_p = 0.1$ and a Spatially Varying Initial Temperature Profile. 164
6.30 Temperature Profiles for $0.1 < Fo_H < 1$ in a Core Wall with $N_p = 0.1$ and a Spatially Varying Initial Temperature Profile. . . 164
6.31 Reductions in Core Heat-Transfer Performance for a Core with a Spatially Varying Initial Temperature Profile and $N_p = 0.1$. . . . 165
6.32 Responses of the Outlet Temperature of the Fluid Flow to Single Flow Exposures of a Regenerator-Core Passsage. Reprinted from Beck [5] with permission from the ASME. 169
6.33 Normalized Effectiveness Responses for $C_{RAT} = (hA)' = 1$; $NTU = 1, 5, 10, 20$, and 40. Reprinted from Beck [5] with permission from the ASME. 171
6.34 Normalized Effectiveness Responses for $(hA)' = 1$; $NTU = 20$; $C_{RAT} = 0.1, 0.5$, and 1. Reprinted from Beck [5] with permission from the ASME. 171

6.35 Schematic of an Experimental Apparatus for Tests of Heat-Transfer Performance. Reprinted from Beck [5] with permission from the ASME. 174
6.36 End-View Enhanced Pictures of Magnified Core Specimens. Reprinted from Beck [5] with permission from the ASME. 177
6.37 Response of Core Specimen 1 to a Single Flow Exposure. Reprinted from Beck [5] with permission from the ASME. 178
6.38 All Effectiveness-Response Data from Runs 1–12. Reprinted from Beck [5] with permission from the ASME. 179
6.39 Axial-View Schematic of an Axial-Flow Rotary-Regenerator Core with Either Uniform-Width Seals (Solid Lines and Crosshatched area) or Sector Seals (Dashed Lines). 181
6.40 Definitions of Δ_1 and Δ_2. 182
6.41 Definition of $\phi(r)$. 182
6.42 (a) Relevant Areas Used to Calculate Δ_1 and (b) Relevant Areas Used to Calculate Δ_2. 183
6.43 Schematics of Two Example Regenerators with Extreme Configurations, Demonstrating the Effect of Varying w/R_o: (a) $w/R_o \to 0$ (b) $\phi(R_i/w) = \theta/2$. 190
6.44 Schematics of Two Example Regenerators with Extreme R_i/R_o Values: (a) $R_i/R_o \to 1$ (b) $R_i/R_o \to 0$. 191
6.45 Schematics of Example Regenerators for which Constraints are Violated. 192
6.46 Schematics of Regenerators with Different Solid-Area Ratios. . . 192
6.47 Plots of Δ_W vs. w/R_o for $NTU = 22$; $C_{RAT} = 1$; $(hA)' = 1$; $C_{ROT} \to \infty$; $As' = 0.1$; and $R_i/R_o = 0.1, 0.3, 0.5, 0.7$, and 0.9. . 193
6.48 Plot of Δ_W vs. w/R_o for $NTU = 22$; $C_{RAT} = 1$; $(hA)' = 1$; $C_{ROT} \to \infty$; $As' = 0.5$; and $R_i/R_o = 0.1, 0.3, 0.5, 0.7$, and 0.9. . 194
6.49 Plot of Δ_W vs. w/R_o for $NTU = 22$; $C_{RAT} = 1$; $(hA)' = 1$; $C_{ROT} \to \infty$; $As' = 1.0475$; and $R_i/R_o = 0.1, 0.3, 0.5, 0.7$, and 0.9. 194
6.50 Plots of ϵ'_W and ϵ vs. r/w for $C_{RAT} = 1$; $(hA)' = 1$; $C_{ROT} \to \infty$; $As' = 0.95465$; $R_i/R_o = 0.5$; $w/R_o = 0.9$; and $NTU = 22$, 50, and 100. 195
6.51 Plots of Δ_W vs. w/R_o for $C_{RAT} = 1$; $(hA)' = 1$; $As' = 1.0475$ or 0.95465; $R_i/R_o = 0.5$; and $NTU = 22$, 50, and 100. 196
6.52 Plot of ϵ vs. C_{RAT} for $NTU = 22.0$; and $C_{ROT} \to \infty$ 196
6.53 Plot of $\Delta C_{RAT}/C_{RAT}$ vs. r/w for $NTU = 22$; $(hA)' = 1$; $C_{ROT} \to \infty$; $As' = 0.95465$; $R_i/R_o = 0.5$; and $w/R_o = 0.9$. 197
6.54 Plots of Δ_W vs. w/R_o for $NTU = 22$; $(hA)' = 1$; $C_{ROT} \to \infty$; $As' = 0.95465$; $R_i/R_o = 0.5$; and $C_{RAT} = 0.7, 0.8, 0.9$, and 1. . . 198
6.55 Schematic of a Regenerator with $As' = 0.5$; $R_i/R_o = 0.5$; and $w/R_o = 0.6$. 199
6.56 Schematic of a Two-Chamber Switching Regenerator. 203
6.57 θ Definition. 204
6.58 Example Regenerator Core with $\alpha_{ROT} = 45^o$ and $As' = 1$. 206
6.59 Example Regenerator Core with $\alpha_{ROT} = 45^o$ and $\Delta = 15^o$. . . . 207

6.60 Example Regenerator Core Modelled as Two Separate Cores. . . 208
6.61 Division of α_{ROT} Core-Portion into N_{ROT} Subportions for a Discontinuous-Rotation Regenerator with $\alpha_{ROT} = 54^o$ and $\Delta = 6^o$. 212
6.62 Propagation of Core Subportion for a Discontinuous-Rotation Regenerator with $\alpha_{ROT} = 54^o$ and $\Delta = 6^o$. 213
6.63 Schematic of a Regenerator Core with a Discontinuous-Rotation Angle, $\alpha_{ROT} = 120^o$ and Subregenerator Cores for $\Delta = -20^o$ and $\Delta = 30^o$. 215
6.64 ζ_j vs. $-60^o \leq \Delta \leq 60^o$ for $\alpha_{ROT} = 120^o$. 216
6.65 C_N-Side Outlet Temperatures vs. Δ for Each Subregenerator of a Discontinuous-Rotation Regenerator with $\alpha_{ROT} = 120^o$, $C_{RAT} = (hA)' = 1$, and $NTU = 0.1$. 216
6.66 Δ_{DR} vs. Δ for a Discontinuous-Rotation Regenerator with $\alpha_{ROT} = 120^o$, $C_{RAT} = (hA)' = 1$, and $NTU = 0.1$. 217
6.67 C_N-Side Outlet Temperatures for Each Subregenerator of a Discontinuous-Rotation Regenerator with $\alpha_{ROT} = 120^o$; $C_{RAT} = (hA)' = 1$; and $NTU = 10$. 218
6.68 Δ_{DR} vs. Δ for a Discontinuous-Rotation Regenerator with $\alpha_{ROT} = 120^o$; $C_{RAT} = (hA)' = 1$; and $NTU = 10$. 219
6.69 C_N-Side Outlet Temperatures vs. Δ for Discontinuous-Rotation Regenerators with $\alpha_{ROT} = 120^o$; $(hA)' = 1$; $NTU = 20$; and $C_{RAT} = 1$ and $C_{RAT} = 0.7$. 219
6.70 Δ_{DR} vs. $-60 < \Delta < 60$ for Discontinuous-Rotation Regenerators with $\alpha_{ROT} = 120^o$; $(hA)' = 1$; $NTU = 20$; and Various C_{RAT} Values. 220
6.71 Δ_{DR} vs. Δ for Discontinuous-Rotation Regenerators with $\alpha_{ROT} = 120^o$; $(hA)' = 1$; $NTU = 0.1$; and $C_{RAT} = 1$ and $C_{RAT} \to 0$. . . . 221
6.72 Δ_{DR} vs. Δ for Discontinuous-Rotation Regenerators with $\alpha_{ROT} = 120^o$; $C_{RAT} = 1$; $(hA)' = 0$, 1, and ∞; and $NTU = 10$. 221
6.73 Δ_{DR} vs. Δ for Discontinuous-Rotation Regenerators with $\alpha_{ROT} = 120^o$; $C_{RAT} = 1$; $(hA)' = 0$, 1, and ∞; and $NTU = 5$. 222
6.74 Δ_{DR} vs. Δ for a Discontinuous-Rotation Regenerator with $\alpha_{ROT} = 120^o$; $C_{RAT} = 1$; $(hA)' = 0$, 1, and ∞; and $NTU = 1$. 222
6.75 Δ_{DR} vs. Δ for Discontinuous-Rotation Regenerators with $C_{RAT} = (hA)' = 1$; $NTU = 10$ and 25; and $\alpha_{ROT} = 45^o$. 223
6.76 Δ_{DR} vs. Δ for Discontinuous-Rotation Regenerators with $C_{RAT} = (hA)' = 1$; $NTU = 10$ and 25; and $\alpha_{ROT} = 30^o$. 224
6.77 Δ_{DR} vs. Δ for Discontinuous-Rotation Regenerators with $C_{RAT} = (hA)' = 1$; $NTU = 10$ and 25; and $\alpha_{ROT} = 20^o$. 224
6.78 Seal Leakage Calculated Using the Method of Harper [20]. 228
6.79 Schematic of a Regenerator-Core Cross Section with Inlet and Outlet Headers. 230

Gas-Turbine Regenerators

Chapter 1

Introduction

This chapter introduces gas-turbine regenerators. This chapter describes all aspects of gas-turbine regenerators that are considered in this book. It also defines system parameters that are used to quantify the performance of a gas-turbine regenerator.

Regenerative gas turbines employ either of two kinds of heat exchangers: (1) regenerators or (2) recuperators. Section 1.1 contrasts regenerators with recuperators. Section 1.1 shows that regenerators have advantages over recuperators for regenerative gas turbines. Regenerator performance consists of heat transfer, leakage, pressure drops, and power consumption. Sections 1.2–1.5 introduce these performances.

1.1 Regenerators vs. Recuperators

Figure 1.1 shows two applications of heat exchangers that can improve the performances of gas turbines:

1. Exhaust-gas heat exchange; and

2. Intercooling.

The exhaust-gas heat exchanger recovers heat from the exhaust and uses it to preheat the compressed air before the compressed air enters the combustor. Preheating of the compressed air premits a small fuel input to the combustor for a given power output of the engine and, therefore, a high efficiency. The intercooler absorbs heat from the compressed air between the compressor stages. Less power is required to compress cool gas than is required to compress warm gas, so the intercooler reduces the power consumed by the high-pressure compressor (HPC) and increases the net power output of the engine.

Figures 1.2 and 1.3 show two kinds of heat exchangers that can be used for exhaust-gas heat exchange and intercooling. Figure 1.2 is a schematic of

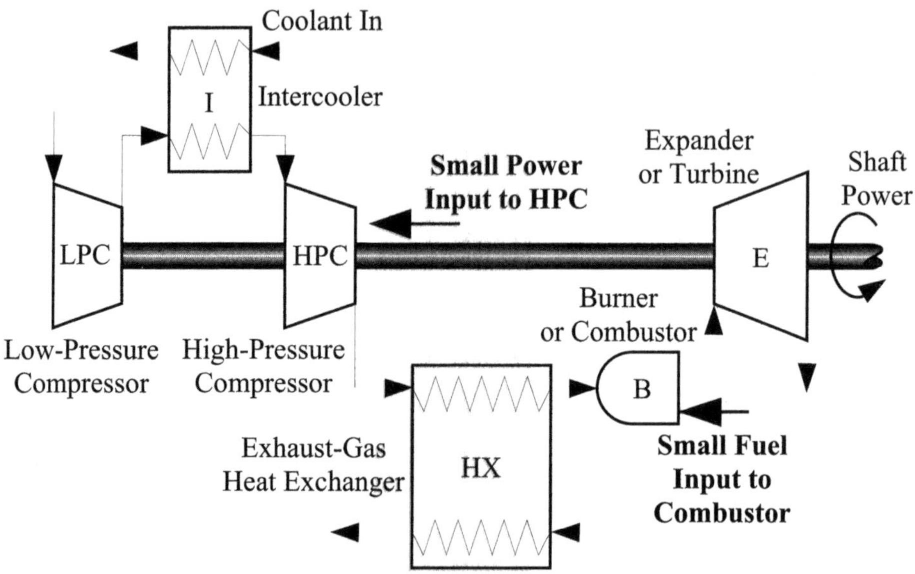

Figure 1.1: How Heat Exchangers Improve the Performances of Gas Turbines.

an axial-flow rotary-regenerator core, and Figure 1.3 is a schematic of a recuperator core[1]. As heat exchangers, the principal functions of regenerators and recuperators are to transfer heat between two fluid flows. In a regenerator, the same heat-transfer surface is alternately exposed to the flows. In a recuperator, heat is transferred through walls that separate the flows[2].

Regenerators have at least three *advantages* over recuperators for regenerative gas turbines (see Kays and London [37]).

1. Regenerators can be made more compact than recuperators.

2. Regenerators can be made with lower material and manufacturing costs than recuperators.

3. Regenerators have fewer fouling problems than recuperators, due to the periodic flow reversals through regenerator cores.

Regenerators have two *disadvantages* compared to recuperators:

1. Regenerators have seal leakage, which reduces the thermal efficiencies of the gas turbines that contain them.

[1] Regenerators and recuperators typically contain headers, which distribute the flows over the core. For clarity, Figures 1.2 and 1.3 do not show headers.

[2] Kays and London [37] provide comprehensive information on recuperators, in addition to basic information on regenerators.

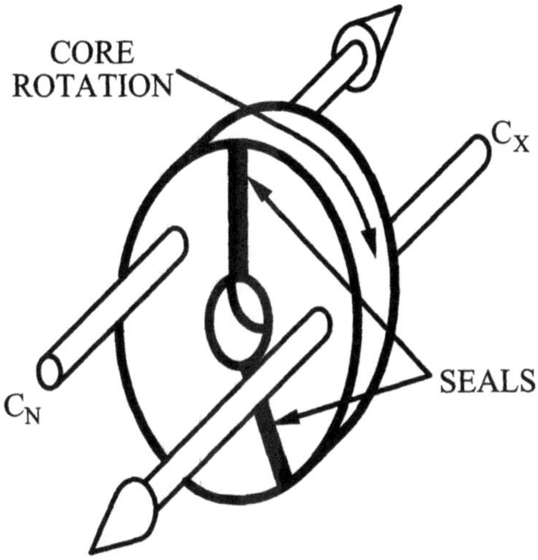

Figure 1.2: Schematic of an Axial-Flow Rotary-Regenerator Core.

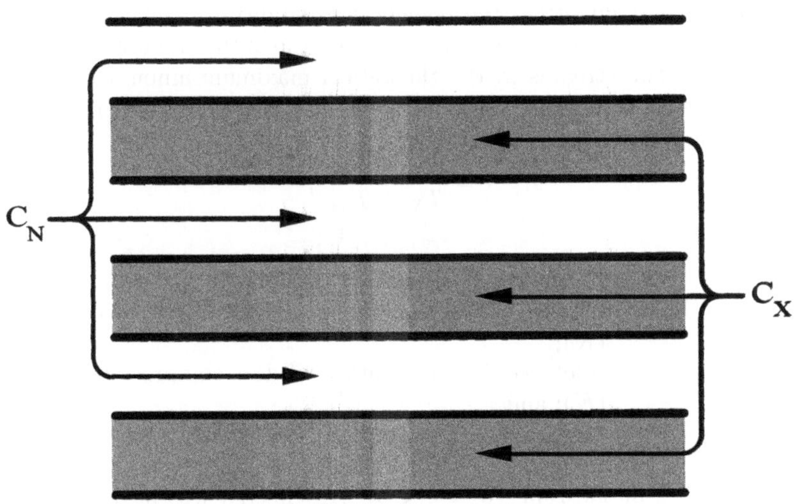

Figure 1.3: Schematic of a Counter-Flow Recuperator Core.

2. Unlike recuperators, regenerators are not passive—regenerators require power inputs to move their cores from one flow to the other[3].

The advantages associated with regenerators typically out-weigh the seal-leakage and power-consumption problems, however. In addition, regenerators can be designed for negligible seal leakage and negligible power consumption (see Chapters 4 and 5). This book focuses on regenerators (although some of the provided results apply for recuperators as well as regenerators).

The remaining sections of this chapter introduce the four kinds of regenerator performance: (1) heat transfer; (2) leakage; (3) pressure drops; and (4) power consumption. To introduce these four kinds of regenerator performance we focus on gas-turbine regenerators for exhaust-gas heat exchange, because this is the most common application of regenerators in gas turbines; however, regenerators can also be used for intercooling.

1.2 Heat Transfer

1.2.1 Regenerator Effectiveness

For a regenerative gas turbine to achieve high performance, its regenerator must transfer a large amount of heat from the exhaust to the compressed air. Figure 1.4 shows temperatures through a regenerator core used for exhaust-gas heat exchange. For the regenerative gas turbine to have a high efficiency, the regenerator should heat the compressed air to a temperature close to the turbine-exit temperature so that the fuel input to the combustor will be minimized. Regenerator effectiveness quantifies the regenerator's ability to heat the compressed air. Regenerator effectiveness is defined as the ratio of the amount of heat transferred between two streams to the theoretical maximum amount of heat that could be transferred by a perfect heat exchanger. For gas-turbine regenerators, this definition reduces to:

$$\epsilon = \frac{T_{No} - T_{Ni}}{T_{Xi} - T_{Ni}}, \tag{1.1}$$

where

ϵ	=	Regenerator effectiveness (–);
T_{No}	=	Outlet (warm) temperature of the compressed air (K);
T_{Ni}	=	Inlet (cool) temperature of the compressed air (K); and
T_{Xi}	=	Inlet (hot) temperature of the exhaust (K).

Effectiveness lies in the range $0 \leq \epsilon < 1$. An effectiveness of 0 is associated with no heat exchanger at all ($T_{No} = T_{Ni}$). A perfect effectiveness of 1 is unachievable because it is impossible to heat the compressed air to exactly the

[3] In a switching regenerator, the cores are stationary, but the flows are switched (which requires some power input).

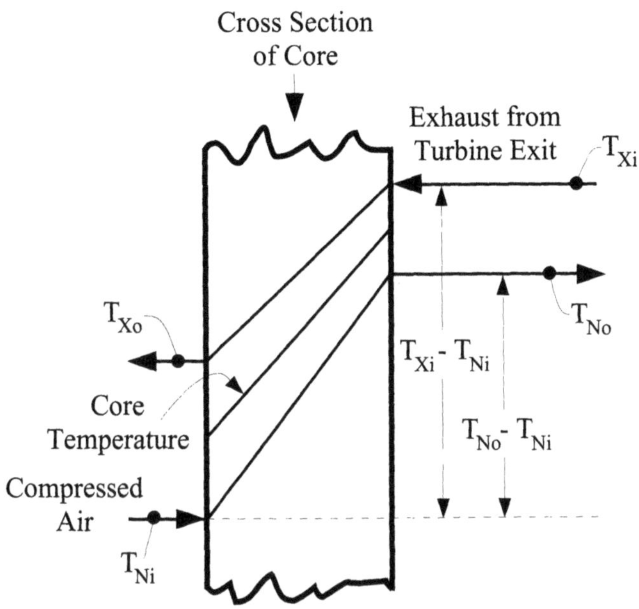

Figure 1.4: Temperatures through a Regenerator Core.

turbine-exit temperature (T_{Xi}). Effectivenesses of gas-turbine regenerators are typically in the range $0.8 \leq \epsilon \leq 0.95$.

This section has quantified heat transfer with effectiveness. The next section shows the effect of the heat-capacity rates of the flows on regenerator effectiveness.

1.2.2 Heat-Capacity Rates

The principal mode of heat transfer in an exhaust-gas regenerator is convective heat transfer from the exhaust to the core and from the core to the compressed air. From Newton's law of convective heat transfer the heat-transfer rate from the exhaust to the core (and then from the core to the compressed air) is greatest if the temperature differences are large. The temperature differences are largest if the heat-capacity rate (product of mass-flow rate, $\dot{m}$, and specific heat capacity at constant pressure, c_P) of the exhaust flow is much greater than the heat-capacity rate of the compressed-air flow. If this is true, then the temperature of the exhaust does not change significantly, but the temperature of the compressed air changes drastically, and the compressed air can be heated to a temperature almost equal to the turbine-exit temperature (T_{Xi})[4]. This argument indicates that regenerator effectiveness depends on the relative sizes of the heat-capacity

[4]For intercoolers, the heat-capacity rate of the coolant flow is usually made much greater than that of the compressed-air flow because (1) for a fixed intercooler size, effectiveness is maximized (the compressed-air flow is cooled to the greatest extent) and (2) the pumping power required to achieve the high heat-capacity rate of the coolant is usually low because

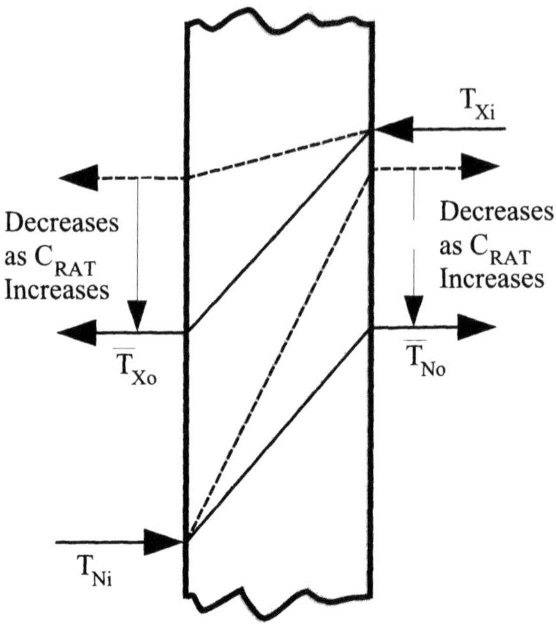

Figure 1.5: Effect of Heat-Capacity-Rate Ratio (C_{RAT}) on Regenerator Effectiveness. Reprinted from Beck [7] with permission from the ASME.

rates of the flows, or the ratio of the heat-capacity rates. The heat-capacity-rate ratio is defined as

$$C_{RAT} \equiv \frac{C_N}{C_X} , \tag{1.2}$$

where

C_{RAT}	=	Heat-capacity-rate ratio (–);
C_N	=	Heat-capacity rate of the compressed-air flow (W/K); and
C_X	=	Heat-capacity rate of the exhaust flow (W/K).

If we hold everything else constant and reduce the heat-capacity-rate ratio in a regenerator, we can expect an increase in effectiveness (ϵ). For example, if we took an operating regenerator and ducted some of the compressed air around the core, then the heat-capacity rate of the compressed air would be reduced, the heat-capacity-rate ratio would be reduced, and effectiveness would be increased—the smaller flow of compressed air passing through the core would be heated to a higher temperature. Figure 1.5 shows the effect of the heat-capacity-rate ratio on regenerator effectiveness.

So far, we have quantified the heat transfer in a regenerator with effectiveness (ϵ), and shown the effect of the flow heat-capacity rates (C_N and C_X) and the

liquid coolants (with low kinematic viscosities) are usually used.

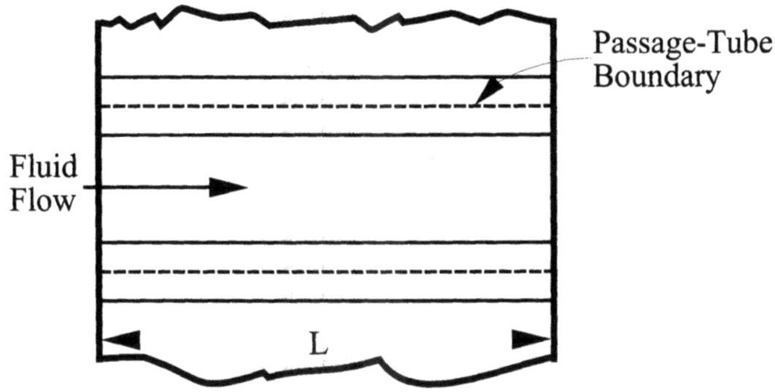

Figure 1.6: Schematic of an Isolated Regenerator Passage Tube (The ratio, D_H/L, is much larger than what is typical for gas turbines). Reprinted from Beck [5] with permission from the ASME.

ratio of these heat-capacity rates (C_{RAT}) on effectiveness. The next section considers how the regenerator core aborbs heat from the exhaust through a convective conductance, $(hA)_X$, and releases this heat to the compressed air through a convective conductance, $(hA)_N$.

1.2.3 Convective Conductances and Number of Transfer Units

A regenerator core consists of many passages that are alternately exposed to the compressed air and the exhaust. When considering heat transfer in a regenerator, it is convenient to conceptually isolate a single passage and its associated passage walls, as shown in Figure 1.6. We refer to this passage/passage-wall combination as a passage tube. When the passage tubes are on the exhaust side of the regenerator, they absorb heat through a convective conductance, $(hA)_X$. When they are on the compressed-air side, they release the heat through a convective conductance, $(hA)_N$.

For high heat transfer, the convective conductances of a regenerator must be high (relative to the heat-capacity rates) so that heat can be transferred well to the core from the exhaust and from the core to the compressed air. The resistance to heat transfer from one flow to another is the sum of the convective resistances associated with each flow. Therefore, the overall convective conductance from one flow to the other is

$$(hA)_{N+X} = \frac{1}{1/(hA)_N + 1/(hA)_X} , \tag{1.3}$$

where

$(hA)_{N+X}$	=	Overall convective conductance (W/K);
$(hA)_N$	=	Convective conductance on the compressed-air side (W/K); and
$(hA)_X$	=	Convective conductance on the exhaust side (W/K).

A dimensionless overall convective conductance is known as the Number of Transfer Units:

$$NTU \equiv \frac{1}{C_N[1/(hA)_N + 1/(hA)_X]} , \tag{1.4}$$

where

NTU	=	Number of Transfer Units (–); and
C_N	=	Heat-capacity rate of the compressed-air flow (W/K).

The Number of Transfer Units quantifies the size of a regenerator core in a dimensionless sense[5]. Therefore, as the Number of Transfer Units increases, so does effectiveness.

In addition to the Number of Transfer Units (NTU), the ratio of the convective conductances, $(hA)'$, also affects regenerator heat transfer; however, its influence is usually slight for gas-turbine regenerators. The convective-conductance ratio is defined as

$$(hA)' \equiv \frac{(hA)_N}{(hA)_X} , \tag{1.5}$$

where

$(hA)'$	=	Convective-conductance ratio (–);
$(hA)_N$	=	Convective conductance on the compressed-air side (W/K); and
$(hA)_X$	=	Convective conductance on the exhaust side (W/K).

For heat-capacity-rate ratios greater than 90% ($C_{RAT} > 0.9$), convective-conductance ratios that differ from unity by up to 80% have only a small ($< 1\%$) effect on effectiveness (ϵ). However, for lower heat-capacity-rate ratios ($C_{RAT} \leq 0.9$) such as those used for intercoolers, the effects of $(hA)'$ can be as much as 7% (see Lambertson [40]).

This section has defined the Number of Transfer Units (NTU), which quantifies regenerator size in a dimensionless sense. When we design regenerators, we first calculate the Number of Transfer Units (NTU) required to achieve a specified effectiveness for specified flows, etc., then we convert this dimensionless

[5]For example, the regenerator in a 10 kW hybrid vehicle and the regenerator in a 200 MW power-generation gas turbine both could have NTU= 50; however, the hybrid vehicle's regenerator would fit on a desk top, but the 200 MW gas turbine's regenerator would span the length of a basketball court.

regenerator size into physical core dimensions based on the heat-transfer and pressure-drop performances of the regenerator core. Core Compactness (CC) quantifies the heat-transfer performance of the core, and core permeability (κ) quantifies the pressure-drop performance of the core. Core compactness quantifies the amount of convective conductance a regenerator core has per unit of core volume. The next section defines core compactness. Section 1.4 defines core permeability (κ) in the context of the pressure-drop performance of regenerators.

1.2.4 Core Compactness

As noted in Section 1.1, one advantage of regenerators as heat exchangers is their compactness. The small passage dimensions of regenerator cores enable regenerators to have a high degree of potential for heat-transfer performance per unit of core volume. Core compactness is defined as the amount of convective conductance (potential for heat-transfer performance) per unit of core volume (with the thermal conductivity of the fluid, k, included in the denominator to facilitate comparison of regenerator cores used with different fluids):

$$CC \equiv \frac{(hA)}{kV_R} , \tag{1.6}$$

where

CC	$=$	Core compactness (m^{-2});
(hA)	$=$	Convective conductance (W/K);
k	$=$	Thermal conductivity of the fluid ($W/m-K$); and
V_R	$=$	Volume of the core (m^3).

The flows through gas-turbine regenerators are typically laminar. In addition, uniform passage geometries are typical. Under these conditions, the core compactness can be expressed as

$$CC = \frac{4pNu}{D_H^2} , \tag{1.7}$$

where

p	$=$	Core porosity, the ratio of voids volume to total volume (–);
Nu	$\equiv$	hD_H/k, the appropriate Nusselt number for laminar flow (–);
k	$=$	Thermal conductivity of the fluid ($W/m-K$);
D_H	$\equiv$	$4A_{ff}/\wp$, the hydraulic diameter of a core passage (m);
A_{ff}	$=$	Free-flow area of the passage (m^2); and
$\wp$	$=$	Wetted perimeter of the passage (m).

Equation 1.7 shows an inverse-squared dependence of core compactness (CC) on passage hydraulic diameter (D_H). Regenerator cores can be easily manufactured with much smaller passages than recuperators. This makes the core compactnesses of regenerators much greater than the core compactnesses of recuperators.

The most significant difference between regenerators and recuperators from a heat-transfer standpoint is that the cores of regenerators rotate. The next section shows how core rotation affects regenerator effectiveness.

1.2.5 Core Rotation

The rotation of rotary regenerators affects regenerator heat transfer (effectiveness, ϵ). Consider the passage tubes of a regenerator as it rotates. During the exposure to the exhaust flow, heat is convected from the exhaust to the passage tubes. During the exposure to the compressed-air flow, the heat is convected from the passage tubes to the compressed-air flow. From Newton's law of convection, convective heat transfer is greatest when the temperature difference between the surface and the fluid is greatest. For greatest heat-transfer rates in a regenerator, the passage tubes should not be permitted to be heated or cooled to the temperatures of the flows, so they should not be permitted to remain exposed to either flow for long time periods. Short time periods of flow exposure are effected with fast core-rotation rates. This section quantifies "short" and "long" time periods.

The heat-capacity rate of the core material, C_R, quantifies a regenerator's core-rotation rate. A dimensionless core-rotation rate is

$$C_{ROT} \equiv \frac{C_R}{C_N} , \tag{1.8}$$

where

C_{ROT}	=	Dimensionless core-rotation rate (–);
C_R	=	Heat-capacity rate of the core (W/K);
C_N	=	Heat-capacity rate of the compressed air (W/K).

From the above argument, regenerator effectiveness is highest for fast core-rotation rates and reduced for slower rotation rates. Typically designers of regenerative gas turbines keep the dimensionless core-rotation rate in the range $1 \leq C_{ROT} \leq 5$. A "short" time period of exposure to the flows is associated with $C_{ROT} \geq 5$. A "long" time period is associated with $C_{ROT} \leq 2$. (See Chapters 5 and 6.)

1.2.6 Core Conduction

Conduction in regenerator cores in the direction of the flows reduces the heat transfer of the regenerators (their effectivenesses). To illustrate the effects of core conduction, we consider a fixed value of the heat-capacity-rate ratio ($C_{RAT} = 1$) in the following sections. In addition, we assume that the core

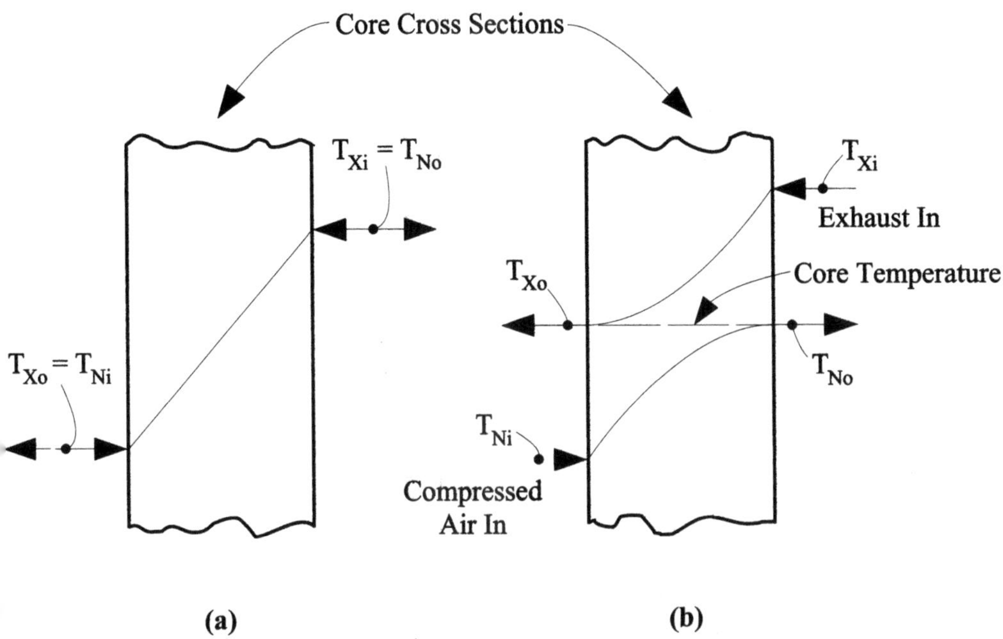

Figure 1.7: Temperatures through a Regenerator Core with Equal Heat-Capacity Rates ($C_{RAT} = 1$) and (a) a Perfect Effectiveness ($\epsilon \to 1$) and No Axial Conduction and (b) Infinite Conduction.

rotates sufficiently fast that the temperature profile through the core does not change significantly during each rotation.

Core Conduction During Flow Exposures

First consider exposures of the core to the flows. Figure 1.7 (a) shows temperatures through a regenerator core for equal heat-capacity rates ($C_{RAT} = 1$), a perfect effectiveness ($\epsilon \to 1$), and no conduction. The high effectiveness requires a steep core temperature profile. With axial conduction in the core, heat flows from the warm face of the core to the cooler face. Thus, the temperature profile is more shallow with axial conduction in the core than with no axial conduction. Figure 1.7 (b) shows temperatures through a regenerator core for equal heat-capacity rates ($C_{RAT} = 1$) and infinite core conduction. For infinite core conduction, the temperature of the core is the average of the inlet temperatures of the compressed air and the exhaust. The effectiveness of this regenerator is significantly less than the effectiveness associated with the core shown in Figure 1.7 (a).

In the following, we describe parameters that are used to quantify axial-conduction effects. Axial conduction in regenerators is given by Fourier's law of heat conduction:

$$\dot{Q}_{axial} = -\left(\frac{k_R As_{N,X}}{L}\right)\frac{\partial T_R}{\partial \zeta}\ , \tag{1.9}$$

where

$\dot{Q}_{axial}$ = Axial conduction (W);
k_R = Thermal conductivity of the core material ($W/m-K$);
L = Core thickness (m);
ζ = Dimensionless coordinate in the direction of the flows (–);
T_R = Temperature of the core (K); and
$As_{N,X}$ = Solid areas available for axial conduction (m^2).

Some dimensionless parameters that are used to correlate the effects of axial conduction are the solid-area ratio,

$$As' \equiv \frac{As_N}{As_X}\ , \tag{1.10}$$

and a core-conduction parameter,

$$\lambda \equiv \frac{k_R\,(As_N + As_X)}{C_N L}\ . \tag{1.11}$$

Axial conduction increases with the quantity $k_R(As_N + As_X)/L$, so regenerator effectiveness is decreased to the greatest extent for large values of the core-conduction parameter, λ.

A simple heat balance shows how regenerator effectiveness varies with the the core-conduction paramenter, λ. (Kays and London [37] give a similar argument.) Regenerator effectiveness (ϵ) is defined as the ratio of the amount of heat transferred between two flows to the theoretical maximum that could be transferred with a perfect heat exchanger. Consider a perfect regenerator with an effectiveness of one ($\epsilon = 1.0$) and equal flow heat-capacity rates ($C_N = C_X$)[6]. This regenerator heats the compressed air to exactly the exhaust inlet temperature (T_{Xi}). Also, because the heat-capacity rates are equal, the exhaust flow is cooled to exactly the temperature at which the compressed air enters the core (T_{Ni}). The heat transferred from the exhaust to the compressed air in this perfect heat exchanger is the maximum that can be transferred:

$$\dot{Q}_{MAX} = C_N\,(T_{Xi} - T_{Ni})\ , \tag{1.12}$$

where

$\dot{Q}_{MAX}$ = Maximum heat that can be transferred (W);
C_N = Heat-capacity rate of the compressed air (W/K);
T_{Xi} = Exhaust inlet temperature (K); and
T_{Ni} = Compressed-air inlet temperature (K).

[6]Note that regenerators for automotive gas turbines typically have $\epsilon \approx 0.95$ and $C_N \approx 0.98C_X$.

The temperature profile through the core is linear between the hot face (at a temperature equal to the exhaust inlet temperature, T_{Xi}) and the cool face (at the compressed-air inlet temperature, T_{Ni}). If there were axial conduction through the core, from Fourier's Law of Conduction it would be

$$\dot{Q}_{COND} = \frac{k_R\,(As_N + As_X)\,(T_{Xi} - T_{Ni})}{L} \,, \tag{1.13}$$

where

$\dot{Q}_{COND}$	=	Axial conduction through the core (W);
k_R	=	Thermal conductivity of the core material ($W/m-K$);
As_N	=	Solid face area of the core on the compressed-air side (m^2);
As_X	=	Solid face area of the core on the exhaust side (m^2); and
L	=	Core thickness (m).

This axial conduction would reduce the amount of heat transferred between the flows:

$$\dot{Q}_{ACTUAL} = \dot{Q}_{MAX} - \dot{Q}_{COND} \,, \tag{1.14}$$

where

$\dot{Q}_{ACTUAL}$	=	Actual heat transferred from the exhaust to the compressed air (W).

Thus, the effectiveness of the regenerator would be reduced by

$$\frac{\Delta\epsilon}{\epsilon} \equiv \frac{\epsilon_{IDEAL} - \epsilon_{COND}}{\epsilon_{IDEAL}} = \frac{\dot{Q}_{MAX}/\dot{Q}_{MAX} - \dot{Q}_{ACTUAL}/\dot{Q}_{MAX}}{\dot{Q}_{MAX}/\dot{Q}_{MAX}} = \lambda \,. \tag{1.15}$$

J. R. Mondt [51] found that the effects of axial conduction can be correlated well by the "Mondt number," which is defined as

$$N_{Mo} \equiv \frac{NTU}{\lambda} \,. \tag{1.16}$$

This book correlates the effects of axial conduction with λ alone, instead of with the "Mondt number;" however, the interested reader is referred to Mondt's article [51] and Mondt's discussion of Bahnke and Howard's paper [2].

Core Conduction Under Seals

The last section showed that regenerator effectiveness is reduced if conduction through the core is significant when the core is exposed to the flows. Effectiveness is also reduced due to conduction through the core when the core is under the seals that separate the flows. Heat diffusion causes the temperature profile through the core to degrade toward a mean value. The degree of axial heat diffusion increases with the time period that a passage tube spends under a seal. This time period can be made dimensionless as a Fourier number:

$$Fo_L \equiv \frac{\alpha_R \tau_S}{2L^2}, \tag{1.17}$$

where

Fo_L	=	Fourier number based on core thickness (–);
α_R	=	Thermal diffusivity of the core material (m^2/s);
τ_S	=	Dimensional time period (s); and
L	=	Thickness of the core (m).

Equation 1.17 shows that the effects of axial heat diffusion while passages are under seals are greatest for wide seals and slow core-rotation rates (which produce long time periods, τ_S), high core thermal diffusivities, and thin cores (small L). Seal coverage is a dimensionless parameter that quantifies regenerator seal width:

$$SC \equiv \frac{As_S}{As_N + As_S + As_X}, \tag{1.18}$$

where

SC	=	Seal Coverage, the fraction of the core face area covered by seals (–);
As_S	=	Total solid area available for axial conduction under both of the seals (m^2);
As_N	=	Solid area available for axial conduction on the compressed-air side (m^2); and
As_X	=	Solid area available for axial conduction on the exhaust side.

The Fourier number associated with one seal is

$$Fo_L = \left(\frac{1}{2}\right)\left(\frac{SC}{1-SC}\right)\left(\frac{\lambda}{C_{ROT}}\right), \tag{1.19}$$

where

C_{ROT}	=	Dimensionless core-rotation rate (–); and
λ	=	Dimensionless core-conduction parameter (–).

These past two sections have shown that core conduction in the direction of fluid flows reduces the effectiveness of a regenerator. The next section shows that a high degree of conduction into the walls of the core passages is advantageous.

1.2.7 Porosity

If everything else is held constant, regenerator effectiveness will increase as the porosity of the core is increased. (Note that carry-over leakage also increases with porosity—see Section 1.3—so a trade-off must be made.) Core porosity is the fraction of the core face occupied by passages:

$$p \equiv \frac{A_{ff}}{A_{ff} + As} , \tag{1.20}$$

where

p	=	Core porosity (–);
A_{ff}	=	Free-flow face area of the core (m^2); and
As	=	Solid face area of the core (m^2).

The expression for core compactness (CC) given by Equation 1.7 indicates that core-passage hydraulic diameter should be small to achieve a compact core. We will show in Section 1.4 that small hydraulic diameters have no effect on the total face area of the regenerator core, but they reduce core thickness (with an inverse-squared relationship) and therefore core volume. Porosity should be high to reduce the total face area of the regenerator core, and therefore core volume. Current manufacturing techniques limit hydraulic diameters for ceramic cores to $D_H \approx 0.5$ mm, but high porosities ($p \approx 0.8$) are achievable. The cores of some regenerators are comprised of separated metallic foil strips, which form parallel-plate passage geometries. The strips can be separated by very small distances for small hydraulic diameters[7]. However, the porosities could be small as well, because lower limits are placed on the thicknesses of the foil strips for durability. The effects of low porosity on heat transfer are typically small for metallic cores, but the effects can be large for ceramic cores. This book considers the effects of low porosity on heat transfer for only the parallel-plate passage geometry; however, the qualitative trends are applicable to other passage geometries.

An examination of an isolated passage tube of a regenerator core with a parallel-plate passage geometry shows how porosity affects heat transfer. Figure 1.8 shows an isolated passage tube. During exposures to the flows, heat is convected to (and from) the fluid/solid interface. From the fluid/solid interface, heat diffuses into (and out of) the walls of the passage tube. The convective resistance from the mixed-mean flow temperature to the fluid/solid interface is one resistance to heat transfer. The diffusive resistance of the walls of the passage tube is another resistance to heat transfer. The convective heat-transfer coefficient is defined as

$$h \equiv \frac{(q/A)_0}{T_m - T_0} , \tag{1.21}$$

where

h	=	Convective heat-transfer coefficient ($W/m^2 - K$);
$(q/A)_0$	=	Heat flux to the fluid/solid interface (W/m^2);
T_m	=	Mixed-mean temperature of the fluid (K);
T_0	=	Temperature at the fluid/solid interface (K).

[7]A lower limit could be placed on the hydraulic diameter to prevent fouling.

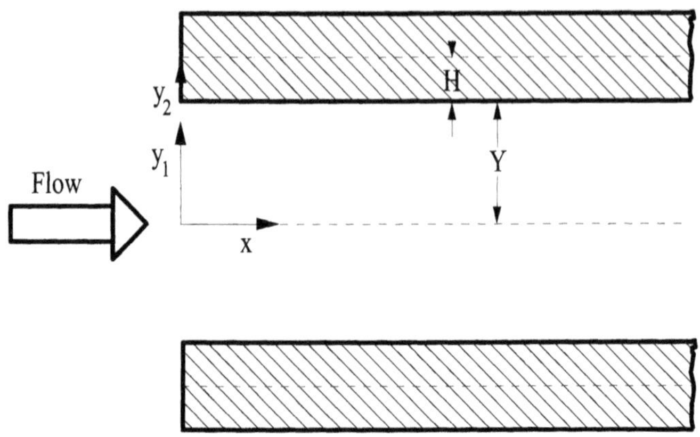

Figure 1.8: Schematic of an Isolated Passage Tube in a Parallel-Plate Regenerator.

We define an effective convective heat-transfer coefficient as

$$h_{ef} \equiv \frac{(q/A)_0}{T_m - \overline{T}_w}, \tag{1.22}$$

where

h_{ef} = Effective convective heat-transfer coefficient $(W/m^2 - K)$;
$\overline{T}_w$ = Average wall temperature (K).

The effective convective heat-transfer coefficient accounts for the diffusive resistance of the walls of the passage tube. The ratio of the two convective heat-transfer coefficients is

$$\frac{h_{ef}}{h} = \frac{T_m - T_0}{T_m - \overline{T}_w}, \tag{1.23}$$

which is less than one. Thus, the ratio of heat-transfer coefficients (h_{ef}/h) can be used to quantify reductions in heat transfer due to low porosity.

We use a porosity number to correlate reductions in heat transfer due to low porosity:

$$N_p \equiv \left(\frac{k}{k_R}\right)\left(\frac{H}{Y}\right), \tag{1.24}$$

where

k = Thermal conductivity of the fluid $(W/m - K)$;
k_R = Thermal conductivity of the core material $(W/m-K)$;
H = One half of the thickness of a passage wall—see Figure 1.8 (m); and
Y = One half of the plate spacing—see Figure 1.8 (m).

For regenerator cores with parallel-plate passage geometries, the porosity number can be expressed as

$$N_p = \left(\frac{k}{k_R}\right)\left(\frac{1-p}{p}\right) , \tag{1.25}$$

where

p = Core porosity (–).

The effects of low porosity are geatest for low conductivity of the core material and (of course) low porosity values. Therefore, effectiveness is low for high porosity numbers.

Section 1.2.6 showed that low core conduction in the direction of fluid flows is required for high effectiveness. This section has shown that high core conduction into the passage walls is required for high effectiveness. In practice, low-conductivity materials (such as ceramics) and high core porosities (thin passage walls) are used to meet these two criteria.

So far we have considered the heat transfer in regenerators only during steady-state operation. The next section introduces the dynamic behavior of regenerators.

1.2.8 Transient Operation

This book considers one kind of transient regenerator operation: transient responses to cold start-ups. Consider a core that initially has a uniform temperature distribution equal to the compressor-delivery temperature (T_{Ni}). Assume that the temperature of the exhaust flow responds with a step function to a value greater than the compressor-delivery temperature: $T_{Xi} > T_{Ni}$. As the core passages first rotate through the exhaust flow they will be heated, but they will not reach the temperatures associated with steady-state operation. Therefore, the outlet temperature on the compressed-air side (T_{No}) and the effectiveness (ϵ) will be less than the values associated with steady-state operation. Transient operation will continue for a time period, τ_{ss}, until the periodic steady-state axial temperature profile of the core is established.

The time period for the axial temperature profile of a regenerator core to come to periodic steady-state is determined principally by the heat capacity of the core, $(mc)_R$, and the larger of the heat-capacity rates of the two fluid flows, C_X. A dimensionless transient time period is

$$\tau^{\star}_{Xss} \equiv \frac{C_X \tau_{ss}}{(mc)_R} . \tag{1.26}$$

where

$\tau^{\star}_{Xss}$ = Dimensionless transient time period (–);
C_X = Heat-capacity rate of the exhaust (W/K);
τ_{ss} = Dimensional transient time period (s); and
$(mc)_R$ = Heat capacity of the core (J/K).

Section 6.1.11 shows that this dimensionless transient time period is of the order unity ($\tau^{\star}_{X\,ss} \sim 1$) for gas-turbine regenerators.

If the flows through the core of a gas-turbine regenerator are non-uniform, then the heat transfer and pressure drops of the regenerator will be affected. The effect on heat transfer is invariably negative. Core pressure drops, on the other hand, can be reduced for non-uniform core-passage geometries. The next section introduces the effects of flow non-uniformity on heat transfer. Section 1.4 introduces the effects on pressure drops.

1.2.9 Flow Non-Uniformity

Flow non-uniformity reduces regenerator effectiveness. Assume one regenerator has uniform flows. Assume a second regenerator is identical in all respects except that the compressed-air flow has a non-uniform distribution over its flow area. The portions of the second regenerator with higher flow rates than the average have lower effectivenesses than the effectiveness of the first regenerator. The portions with lower flow rates than the average have higher effectivenesses than the effectiveness of the first regenerator. The resultant effectiveness of the second regenerator almost always is lower than that of the first, because the high-flow low-effectiveness portions have a greater impact (negative) on the overall effectiveness than the low-flow high-effectiveness portions.

Regenerators may experience flow non-uniformity from four sources:

1. Non-ideal header/core systems[8];
2. Non-uniform core-passage geometries;
3. Seal shape; and
4. Discontinuous rotation.

The following sections describe these four sources of flow non-uniformity.

Non-Ideal Header/Core Systems

Figure 1.9 shows (a) an ideal header/core system and (b) a non-ideal header/core system. An ideal header/core system distributes the flow evenly across the face of the regenerator core. A non-ideal header/core system does not. With an ideal header/core system, all passages of the core are exposed to the same flow history. Therefore, if there is no other source of flow non-uniformity, each passage has the same effectiveness. If a regenerator has non-ideal header/core systems, then the passages of the core can be exposed to different flow histories. Therefore, the effectivenesses of the passages can be different, and the effectiveness of the regenerator will be less than the effectiveness of an otherwise-identical regenerator with ideal header/core systems.

[8] An ideal header/core system produces a uniform flow distribution at its outlet, independent of the flow distribution at its inlet.

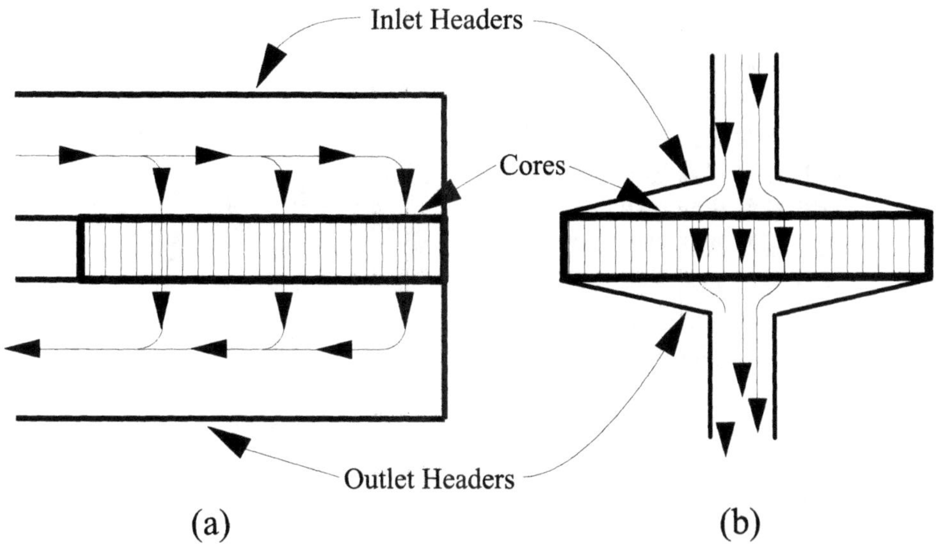

Figure 1.9: (a) An Ideal Header/Core System and (b) a Non-Ideal Header/Core System.

Methods exist for designing nearly-ideal header/core systems that produce nearly-uniform flow distribution (see Wilson [72] and London, et al. [44]). Therefore, this book does not address in detail the effects of flow non-uniformity from non-ideal header/core systems.

Non-Uniform Core-Passage Geometries

Figure 1.10 shows (a) core passages with uniform geometries and (b) core passages with non-uniform geometries. Non-uniform core-passage geometries have a similar effect on heat transfer as do non-ideal header/core systems: the passages can have different effectivenesses, and the effectiveness of the regenerator is usually less than the effectiveness of an otherwise-identical regenerator with uniform passage geometries.

Outdated manufacturing techniques produced regenerator cores with highly non-uniform passage geometries. London [43] addresses the effects of non-uniform core passage geometries on regenerator effectiveness. Current extrusion techniques, however, produce ceramic cores with highly uniform passages. Therefore, this book does not address in detail the effects of flow non-uniformity from non-uniform passage geometries.

Seal Shape

Figure 1.11 shows regenerator seals of alternative shapes. For the annular-sector shaped seals shown in Figure 1.11 (a), passage tubes at every radial location experience the same flow-exposure and seal-coverage histories. For

(a) (b)

Figure 1.10: (a) Core Passages with Uniform Geometries and (b) Core Passages with Non-Uniform Geometries.

seals with shapes different from annular sectors [e.g. the uniform-width seals shown in Figure 1.11 (b)], passage tubes at different radial locations spend different amounts of time exposed to each flow and covered by seals. Therefore, effectiveness is a function of radial location for seals of shapes other than the annular-sector shape. For regenerators with seals of equal area but arbitrary shape, effectiveness is highest if seals of an annular-sector shape are used. The only seal shapes considered in this book are annular-sector shapes and uniform-width shapes; however, the methods given are generally applicable.

Reductions in effectiveness due to seal shape depend on two parameters for uniform-width seals, w/R_o and R_i/R_o. The parameter w/R_o is the ratio of seal width to the outer radius of the core's flow area. The parameter R_i/R_o is the ratio of the inner radius of the core's flow area to the outer radius.

Discontinuous Rotation

Incremental movement of the regenerator core with clamping of the seals during the stationary periods has been proposed by Wilson [75, 76] as a method of reducing direct seal leakage. Figure 1.12 indicates a regenerator core that experiences discontinuous rotation. For general discontinuous-rotation angles, α_{ROT}, some portions of the core experience different flow histories than other portions. Therefore, discontinuous rotation can lead to reductions in regenerator effectiveness. In addition to the discontinuous-rotation angle (α_{ROT}) reductions in effectiveness also depend on the angle Δ, which is defined in Figure 1.12.

This section has introduced the heat transfer of gas-turbine regenerators and the parameters that quantify heat transfer. The following sections introduce the other three regenerator performances: leakage (Section 1.3); pressure drops (Section 1.4); and power consumption (Section 1.5).

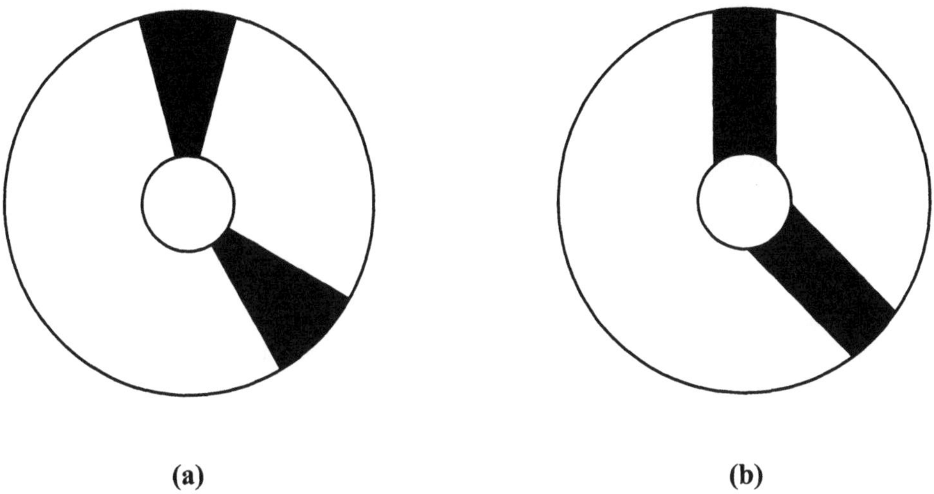

Figure 1.11: Regenerator Seals of Alternative Shapes: (a) Annular-Sector Shaped Seals; and (b) Uniform-Width Seals.

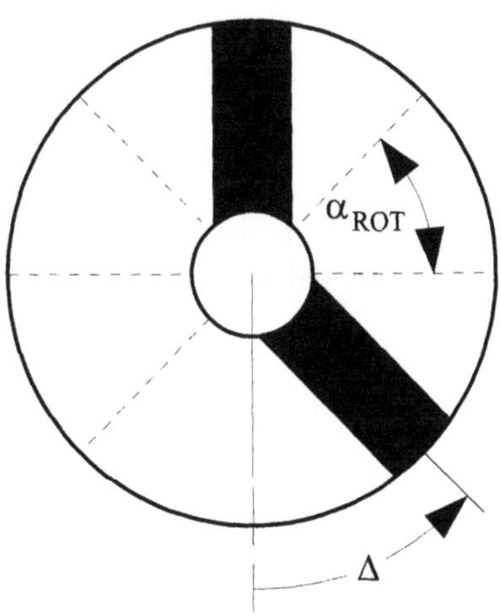

Figure 1.12: Regenerator Core that Experiences Discontinuous Rotation.

1.3 Leakage

Gas-turbine regenerators must have low leakage in addition to high heat transfer. Seal leakage in regenerators has severely limited the performance of gas turbines that include regenerators. For gas-turbine applications, the compressed-air flow is at high pressure. Pumping power is lost when some of the compressed air leaks under the seals to the exhaust-gas flow.

Figure 1.13 shows a typical plot of regenerator leakage as a function of seal clearance. Regenerator seal leakage consists of direct seal leakage and carry-over leakage. Consider the schematic of an axial-flow rotary regenerator shown in Figure 1.2. The fluid flows are at different pressures. Some of the high-pressure fluid leaks directly under the seals to the low-pressure flow. This is the direct seal leakage, $\dot{m}_{XLD}$. There is a second kind of leakage. As the core rotates under seals, some fluid from one flow becomes trapped in the core passages. This trapped fluid is transferred to the other flow when the passages rotate from under the seals. The resulting leakage is the carry-over leakage $\dot{m}_{XLC}$. The sum of the carry-over leakage and the direct seal leakage is the total seal leakage:

$$\dot{m}_{XL\Sigma} = \dot{m}_{XLC} + \dot{m}_{XLD} \tag{1.27}$$

where

$\dot{m}_{XL\Sigma}$ = Total seal leakage (kg/s);
$\dot{m}_{XLC}$ = Carry-over leakage (kg/s); and
$\dot{m}_{XLD}$ = Direct seal leakage (kg/s).

As Figure 1.13 shows, for small seal clearances, the principal contributor to total seal leakage is carry-over leakage. As the seal clearance increases, direct leakage becomes an increasingly greater contributor until it dominates.

Carry-over leakage and direct seal leakage are coupled in one direction (see Harper [26] and Chapter 6.3): carry-over leakage is independent of direct seal leakage, but direct seal leakage depends on the pressure distribution under a seal, and the pressure distribution under a seal is a partial function of carry-over leakage. Therefore, direct seal leakage depends on carry-over leakage. Carry-over leakage can be calculated immediately by

$$\dot{m}_{XLC} = C_{ROT} \left[\frac{p}{1-p}\right] \left[\frac{1}{(\rho c)_R}\right] C_N \rho_c \, , \tag{1.28}$$

where

C_{ROT} = Dimensionless core-rotation rate (–);
p = Core porosity (–);
C_N = Heat-capacity rate of the compressed air (W/K);
$(\rho c)_R$ = Specific heat capacity per unit volume of the core material $(J/m^3 - K)$; and
ρ_c = Density of the carry-over-leakage when it becomes trapped under a seal (kg/m^3).

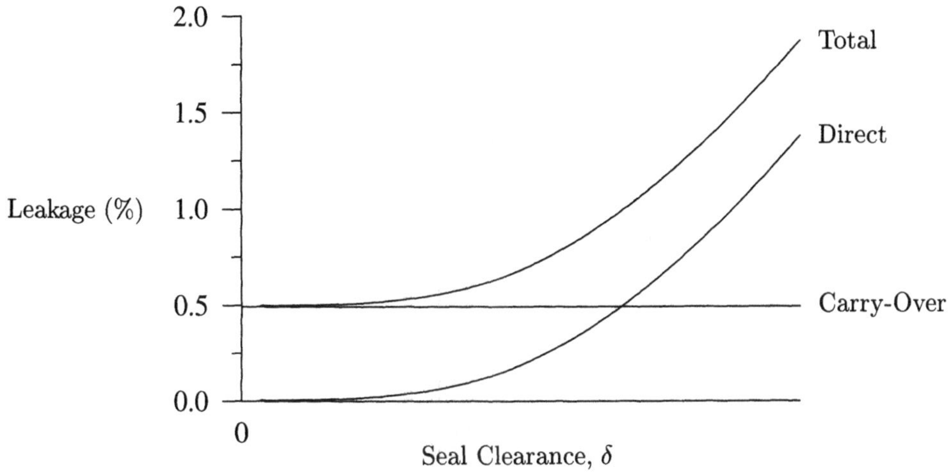

Figure 1.13: Typical Plot of Regenerator Leakage.

Equation 1.28 shows that carry-over leakage increases rapidly with the core porosity (p). However, as mentioned in Section 1.2.7, core porosity should be high for a small total face area of the regenerator core. Therefore, an optimal core porosity must result from a trade-off of increased core face area for reduced carry-over leakage.

Direct seal leakage depends on carry-over leakage and two other parameters. The pressure ratio across the seal, r_R, is one of these parameters. The pressure difference across the seal drives the direct-seal-leakage flow. Therefore, as the pressure ratio increases, so does direct seal leakage and total seal leakage. Direct seal leakage also depends on a dimensionless seal-leakage parameter[9],

$$\left(\frac{BS}{A}\right) \equiv \frac{\dot{m}_{XLC}\sqrt{T_d R n}}{P_c(\alpha\gamma\delta)(R_o - R_i)\sqrt{2}}\,, \tag{1.29}$$

[9]The dimensionless seal-leakage parameter is indicated by BS/A to be consistent with Harper [26].

where

BS/A	=	Dimensionless seal-leakage parameter (–);
T_d	=	Temperature of the direct-seal-leakage fluid (K);
R	=	Gas constant associated with the direct-seal-leakage fluid ($J/kg - K$);
n	=	Number of throttlings under the seal (–);
P_c	=	Pressure of the carry-over-leakage fluid when it becomes trapped under a seal (Pa);
α	=	A flow coefficient (–);
γ	=	A flow coefficient (–);
δ	=	Seal clearance (m);
R_o	=	Outer radius of the core disk (m); and
R_i	=	Inner radius of the core disk (m).

Direct seal leakage typically is greater than carry-over leakage, and (as Figure 1.13 shows) direct seal leakage increases dramatically with the seal clearance. Therefore, typically total seal leakage is low for high values of the seal-leakage parameter BS/A.

1.4 Pressure Drops

In addition to having high heat transfer and low leakage, a regenerator must have low pressure drops, because increased pumping power must compensate for pressure drops across the regenerator.

There is a prevalent misconception about compact heat exchangers. The misconception is that the frontal area of a compact heat exchanger must increase as the passage dimensions are made smaller. In fact, for laminar flows through uniform passages, the frontal area is independent of passage dimension and, therefore, the degree of core compactness:

$$A_{TOT} \neq f(D_H) \; , \tag{1.30}$$

where

A_{TOT}	=	Total frontal area (m^2); and
D_H	=	Hydraulic diameter of the passages (m).

The flow length, on the other hand, *does* vary with passage dimension and the degree of core compactness. Flow length decreases as the inverse-square of passage dimension, and this is what makes the cores of compact heat exchangers so compact:

$$L \propto \frac{1}{D_H^2} \; , \tag{1.31}$$

where

L = Flow length or core thickness (m).

Consider, as an example, two compact heat exchangers (with laminar flows and uniform passages) that have identical pressure drops on the exhaust (or gas-flow) side. If the heat exchangers have identical passage geometries, etc., but Heat Exchanger A has 0.1-mm passages and Heat Exchanger B has 1-mm passages, then the flow areas will be equal, but the core of Heat Exchanger B will be one hundred times as thick as the core of Heat Exchanger A.

Section 1.2.4 defined core compactness, which quantifies the ability of a heat exchanger core to transfer a large amount of heat in a small volume—the degree of compactness of the heat-exchanger core. The analogous parameter that quantifies the pressure-drop performance of a heat exchanger core is permeability:

$$\kappa \equiv \left[\frac{\mu L}{(\Delta P)_{CORE}}\right]\left[\frac{\dot{V}}{A}\right] , \tag{1.32}$$

where

κ	=	Permeability of the core (m^2);
μ	=	Fluid viscosity ($kg/m-s$);
L	=	Thickness of the core (m);
$(\Delta P)_{CORE}$	=	Pressure drop across the core (Pa);
$\dot{V}$	=	Volumetric flow rate through the core (m^3/s); and
A	=	Frontal area exposed to the flow (m^2).

For laminar flow through uniform passages, an alternative expression for the permeability is

$$\kappa = \frac{pD_H^2}{2C_f} , \tag{1.33}$$

where

p	=	Core porosity (–);
D_H	=	Hydraulic diameter of the passages (m); and
C_f	=	Friction coefficient associated with the passage geometry[10] (–).

Equation 1.33 shows that permeability is very low for compact heat exchangers with very small passages (small hydraulic diameters, D_H). Even though the permeability is low for the cores of compact heat exchangers, we achieve low pressure drops across the cores by making the flow lengths (L) short. Fortunately, core compactness varies with the square of the inverse of passage dimension, so we *do not* need to increase the frontal area to give more core volume to compensate for short flow length.

Header pressure drops also contribute to the total pressure drop across the regenerator. Header pressure drop increases with the average velocity at the header inlet, U_i.

This section has introduced pressure drops in regenerators. The final facet of regenerator performance that we consider in this chapter is power consumption, which is introduced in the next section.

1.5 Power Consumption

Regenerators require a finite amount of power to move the core from one flow to the other[11]. Regenerators with clearance seals require a negligible amount of power compared to the power output of the gas turbine. Chapter 4 describes the discontinuous-rotation regenerator and the modular regenerator. Each of these regenerators can require a significant amount of power input to move the core; however, some calculations performed under the supervision of the second author indicate that the power inputs can be low. The discontinuous-rotation regenerator requires a finite amount of power to accelerate the core quickly through the angles of discontinuous rotation. The modular regenerator requires a finite power input to shuffle the core modules. Power consumption is dependent on the specific dimensions, seal technologies, rotation rates, etc. of regenerators. Therefore, this book does not address power consumption in detail.

1.6 Summary

This chapter has introduced the key phenomena that influence the performance of gas-turbine regenerators and the parameters that are used to quantify these phenomena. The next chapter focuses on the historical background of gas-turbine regenerators. Chapter 3 introduces gas-turbine cycles, which are the thermodynamic cycles by which gas turbines generate power from heat. Chapter 4 describes alternative regenerator designs, and Chapter 5 shows how these regenerators can be designed for high performance. Then, the focus returns to regenerator performances in Chapter 6.

---

[11] An exception is the switching regenerator in which the core is stationary, and the flows are switched from one core to the other.

Chapter 2

Background

This chapter briefly describes the history of regenerators and regenerator analysis and design. This chapter sets the context for discussion of state-of-the-art regenerator designs in Chapter 4. First Section 2.1 describes the history of regenerators. Then Section 2.2 describes the history of the ϵ-NTU methods for regenerator analysis and design, as well as other essential analysis and design methods.

2.1 History of Regenerators

We begin with the earliest regenerators and proceed chronologically up to the present, describing landmarks in the development of modern regenerators.

2.1.1 Early Regenerators

We define a regenerator as a device that transfers heat from one fluid to another by exposing a solid or solids alternately to each fluid. Regenerators were pervasive long before they were identified. For example, porous clothing can act as a regenerator core for people in cold climates. Warm exhaled air passes through a scarf (for example) and warms the fibers. Cold ambient air is warmed as it in turn passes through the scarf in the opposite direction. The uncovered lips, the mouth, and the throat also act as regenerator heat-transfer surfaces.

The Romans applied regeneration in a systematic way to heat their baths. They heated stones in the flue of a fire, rolled the hot stones into the water. After the stones had cooled (from heating the water), the Romans removed them and reinserted the stones in the top of the flue. Developments of this overall arrangement using automatically conveyed small pebbles have been patented and used in process furnaces; and they have been considered for gas turbines.

2.1.2 The Stirling Regenerator

Robert Stirling was the first to identify regenerators and to apply them technologically. He applied for his first patent for the Stirling air engine and the regenerator (which he called an economiser) in 1816 (see Walker [71], Zarinchang [86], and Edelman [20]). He also obtained a patent for regenerators as applied to manufacturing processes in 1817. The following quotation of Reese [59] describes Stirling's regenerator for manufacturing processes.

> His invention relates to heating air, gases and fluids by the agency of air, gases or fluids passing through passages formed of brick, metal, stone or any other material adapted to the degree of heat required, and he describes his invention as applicable to the manufacture of glass, pottery and any other case where great heat is required.

2.1.3 The Siemens Regenerator

Friedrich Siemens's patent for a kind of regenerator in 1856 marked the beginning of the modern practical use of regenerators in furnaces (see Reese [59] and the Encyclopaedia Britannica [68]). Hot exhaust gases were ducted through a stack of brickwork containing small passages ("checkerwork") for a period on the order of twenty minutes, and then were switched to pass through a second stack of checkerwork (see Figure 2.1). At the same time air was ducted to the first, now-hot, checkers to pick up heat before being mixed with the fuel and burning over the charged material. His regeneration system was applied to glass-making furnaces a year later, and reduced the heat input 70–80% (see Gale [21]). His brother, Karl Wilhelm (who after being naturalized as a British citizen later became Sir Charles William Siemens), used the system to invent the open-hearth furnace in 1861, having two switchable regenerators. This became predominant in the manufacture of high-quality steel.

2.1.4 Cowper Stoves

The Siemens brothers' work presumably influenced Edward Cowper, who in 1860 patented the hot-blast ("Cowper") stoves to heat the air blown into blast furnaces [15]. These stoves are huge towers (e. g. 10 m in diameter and 40 m high) full of temperature-resistant ceramic checkerwork, heated by the products of combustion of blast-furnace gas for up to two hours, and then switched to take the "cold" blast from turboblowers to produce a hot blast at, nowadays, over 1200 C for typically 40 minutes (see Figure 2.1 and Lankford, et al. [41]). Three or four stoves are used for each blast furnace, one being "on blast" while the others are being heated.

This kind of regenerator has been tried for gas-turbine engines. A serious disadvantage is that immediately after the flow switchover, the compressor must recompress the gas (initially at about atmospheric pressure) in the heated stove.

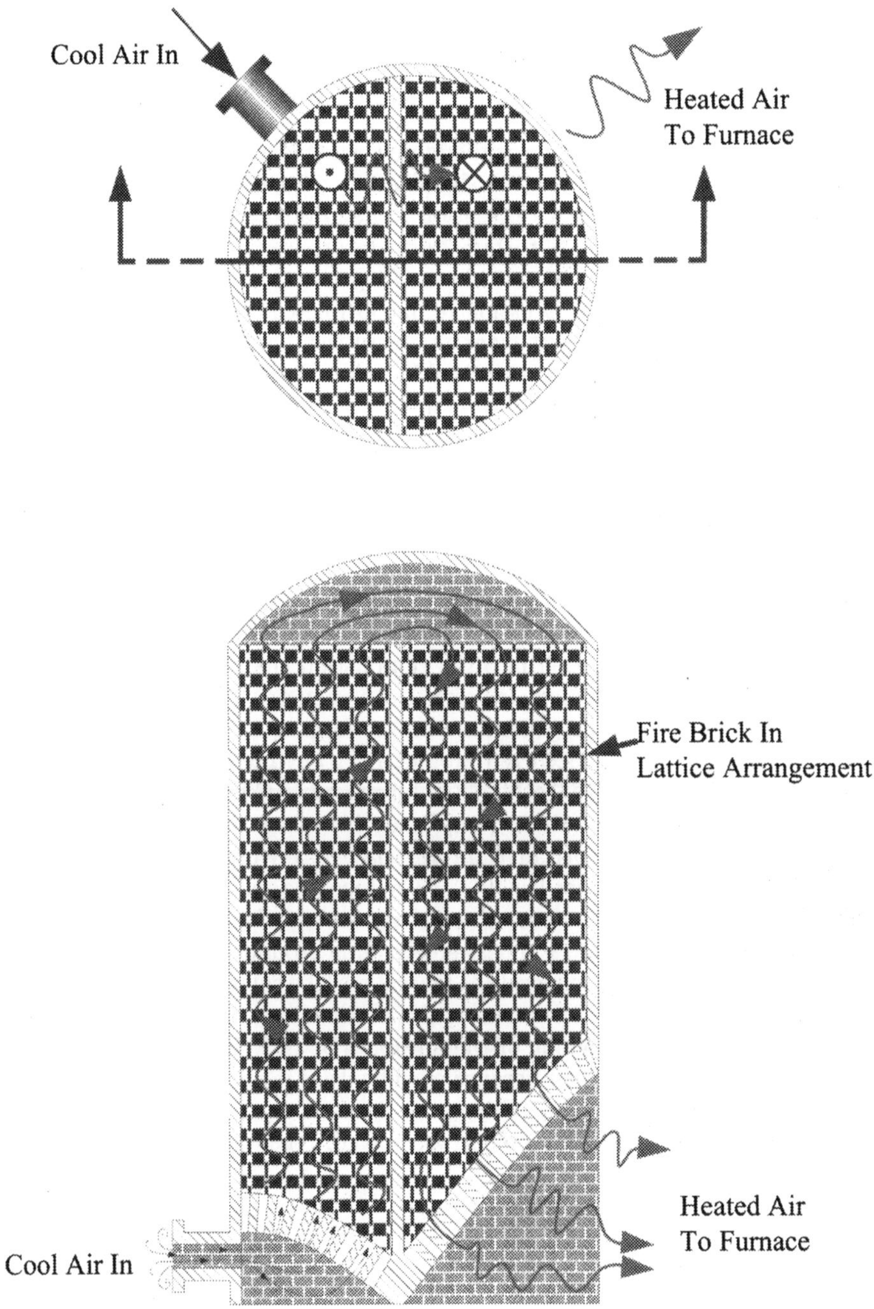

Figure 2.1: Cowper (or Hot) Stove (adapted from Reese [59]).

This process may take minutes, during which time little air is being fed to the combustor and no power is delivered by the turbine.

2.1.5 The Ljungstrom Air-Preheater

A continuous form of regenerator is therefore needed for gas turbines. This need was potentially satisfied by the most significant regenerator development for gas turbines: Fredrik Ljungstrom's invention in 1920 of the rotary air preheater, the precursor of almost all gas-turbine regenerators (see Figure 2.2). We credit Ljungstrom with the first continuous form of regenerator. However, Ljungstrom was likely inspired. In the discussion to Jacob Reese's article on regenerating metallurgic furnaces in the first transactions of the ASME in 1880 [59], a Mr. Holley cites continuous regenerators used by a Mr. Sellers and a Mr. Swinell of Pittsburg (presumably for regenerating metallurgic furnaces).

The air preheater was and is used to transfer heat from the gases leaving the economizer in power-plant steam generators to the incoming furnace air. One effect of the introduction of the air-preheater was to increase boiler efficiency to 88%, giving a fuel saving of 20% [25]. The largest preheaters are now over 10 m in diameter.

The first known application of heat exchangers (in this case fixed-surface recuperators) to gas turbines was by Aegidius Elling in Norway in 1903–4, to raise steam for steam injection into the combustion chambers, and, in a later application, to heat compressed air before the combustor [34] . The first significant known applications of regenerators (in the form of Ljungstrom's rotary air preheater) to gas turbines occurred around 1951 by Huebner at Chrysler [31] and by the UK National Gas-Turbine Establishment (NGTE) [16], in both cases to automobile-size engines. Both groups may have been inspired by Ritz and Hryniszak, who escaped to Britain from Austria before World War II and worked on periodic-flow regenerators [56, 9]. To reduce the leakage, which was potentially very high because of the thermal distortion of matrices produced by winding plain and corrugated stainless-steel ribbon around a hub, the NGTE developed "smart" seals that followed the matrix-surface contours (see Figure 2.3). According to the then NGTE director, Hayne Constant, in a talk to an engineering society attended by one of the authors (D. G. Wilson) in about 1951, the regenerator was developed because of difficulties in increasing the pressure ratio of axial-flow compressors. (Chapter 4 discusses the reasoning behind this statement.) As soon as that problem was solved, regenerator development at the NGTE appears to have ceased.

2.1.6 Chrysler Turbine Cars

Chrysler courageously developed a series of automotive gas turbines using broadly similar stainless-steel matrices for the rotary regenerator, and put fifty cars in the hands of selected members of the public. The principal reason preventing Chrysler from going into production with the turbine car was the high material

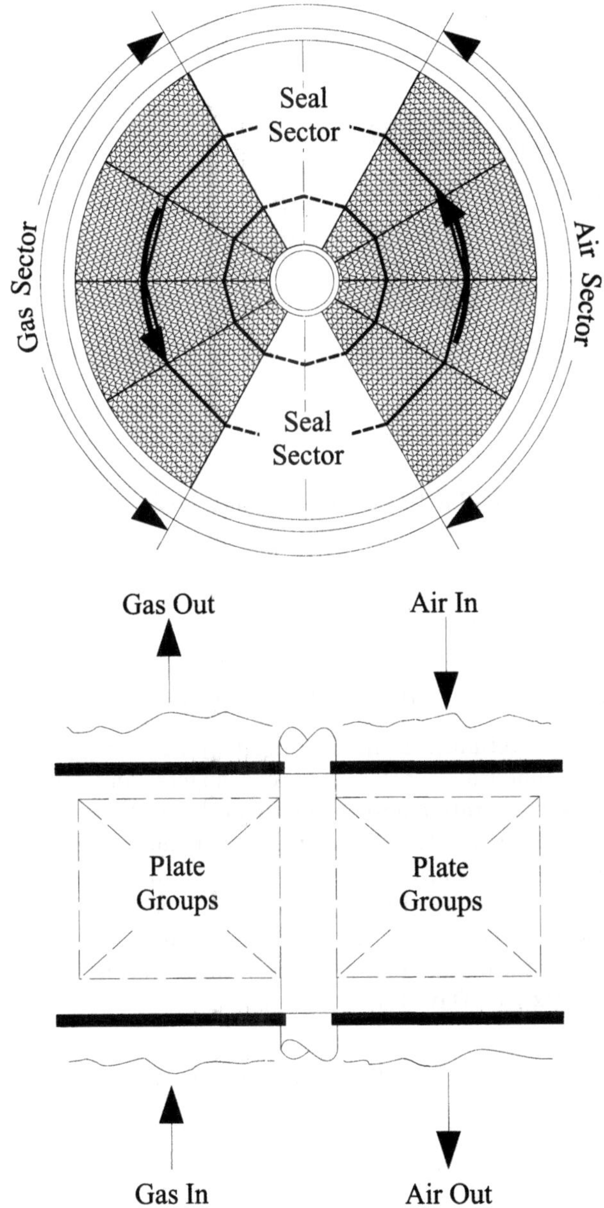

Figure 2.2: Ljungstrom Air Preheater (adapted from Babcock and Wilcox [1] with permission).

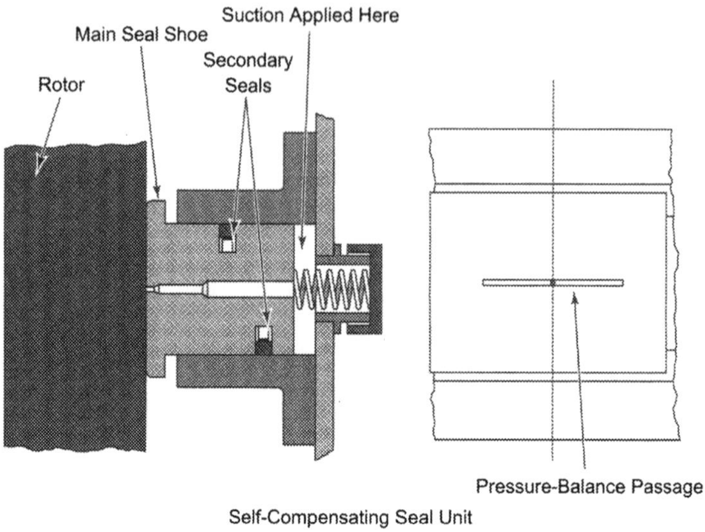

Figure 2.3: NGTE Rotary Regenerator with "Smart" Seals.

cost of the nickel-chromium-cobalt alloys used throughout the engine, including the regenerators (see Huebner [32]).

2.1.7 Chrysler-Corning Collaboration on Ceramics

Chrysler purchased ceramics from Corning to pursue the possibilities of using non-matallic materials for the heat exchangers, and later Chrysler and Corning signed a cooperative agreement to work together to develop ceramic regenerators. Corning developed a method of manufacturing regenerators in high-temperature low-expansion-coefficient ceramic materials (see Lanning and Wardale [42]). The material, in paste form, was coated on two strips of tea-bag paper, one of which was crimped. The plain and wavy ribbons were wound around a rotating mandrel to produce, after firing, a fine "triangular-honeycomb" matrix (see Day [19]).

2.1.8 Rover-Penny Turbine at Le Mans

Corning's development was given world-wide publicity after Noel Penny, then chief gas-turbine engineer at Rover Cars in Britain, incorporated the Corning regenerator into the engine of the Rover-BRM race car competing in the Le Mans 24-hour race (see the magazine Motor [53]). The regenerator earned an outstanding reputation for robustness when the ceramic matrix retained several pieces of blades thrown off by the compressor rotor early in the race, and the decision was made to continue running at reduced power and efficiency without trying to remove the pieces. The car performed amazingly well for a first appearance, not only finishing the race (typically fewer than half do) but finishing in tenth place with an overall fuel consumption of 11.45 miles per US gallon. Without engine damage, the car could achieve a fuel consumption of 22 miles/gallon at a speed of 100 miles/hour [55, 53].

2.1.9 Setback at Ford

The high reputation of the Corning ceramic regenerator dropped precipitously after Ford (US) developed a truck engine (see Cadwell, et al. [10]). Despite apparently thorough testing, production engines experienced regenerator failures because of a reaction of the magnesium-aluminum-silicate ceramic with the sulphur in some high-sulphur Diesel fuels (see Grossman and Lanning [22]). Ford shut down truck-gas-turbine production totally as a result.

Since that time, all known automotive-gas-turbine developments have relied on ceramic rotary regenerators, but many have been plagued by high seal-wear rates and consequent high leakage, up to 15% of the compressor air (see Shimada, et al. [67]).

2.2 Analysis and Design

In this book we use the ϵ-*NTU* methods for regenerator analysis and design. These methods enable the regenerator designer to perform both preliminary (accurate to ±2%) design work and more accurate analysis and design work when a high degree of accuracy is required. In this section we describe how the ϵ-*NTU* methods originated and developed. In addition, we give the history of other analysis and design methods that are essential for designing high-performance gas-turbine regenerators.

2.2.1 Cooling Towers and the *H.T.U.*

The ϵ-*NTU* method of heat-exchanger analysis originated in 1939. London, Mason, and Boelter [45] used the height of a transfer unit, *H.T.U.*, in their analysis of cooling towers. This *H.T.U.* is the height of the cooling tower divided by *NTU*. They defined energy-exchanger and liquid-cooler effectivenesses. Tower performance was expressed in terms of an effectiveness of energy exchanger, as compared to the performance of an infinitely high cooling tower. London and Seban [47] generalized the ϵ-*NTU* method to heat exchangers, mass-transfer systems, and energy-transfer systems (which involve a combination of heat and mass transfer).

2.2.2 Core Data

Regenerator cores that were used during the 1940's consisted of wire mesh or expanded metal or various arrangements of corrugated sheet metal [27]. The heat-transfer and pressure-drop performances of these passage geometries and of passage geometries in contemporary recuperators cannot be predicted well by theory. Also, the theory was not as well developed in the 1940's as it is today. For these reasons, the U. S. Navy Bureau of Ships started a program in 1944 with the goal of experimentally obtaining heat-transfer and fluid-friction design data for compact heat-exchanger cores. First the bureau obtained data for a gas-turbine recuperator/regenerator (terminology from Shah, et al. [64]). In 1947, the Office of Naval Research, in cooperation with the Bureaus of Ships and Aeronautics, extended this work by establishing a similar program at Stanford University. Results of the Stanford program, which lasted until 1971, are data for over one hundred compact-heat-exchanger surfaces (which are reported by Kays and London [37]).

2.2.3 Design Optimization

Harper and Rohsenow [27] addressed the optimization of regenerative gas turbines in a paper published in 1952. They showed the effects of regenerator performance (heat transfer, leakage, and pressure drops) on the performance of the host gas turbine. The theory of regenerator performance was not well enough developed in 1952 for a complete engine optimization; however, Harper later studied leakage performance and effectively completed that part of regenerator-performance theory. His analysis and results were published in 1955 [26].

2.2.4 State of the Art in 1953

Coppage and London [14] summarized the heat-transfer-performance theory that prevailed up to 1953. They indicated mathematical complexities that had inhibited general solution of the governing equations. Also, they gave solutions obtained by numerical-graphical methods under various simplifying assumptions. A notable gap in the theory was lack of exact effectiveness data for finite core-rotation rates.

2.2.5 Finite Core-Rotation Rate

Lambertson made a major contribution to the theory with a paper published in 1958 [40]. He filled the gap in the theory by providing heat-transfer-performance design data in tabular form for finite core rotation and values for heat-transfer parameters that were typical of contemporary regenerator designs. More importantly, however, he demonstrated the usefulness of finite-difference numerical-integration methods and digital computers. Extensions of his method are still used for calculating regenerator effectiveness [7].

2.2.6 Heat Conduction

In a paper published in 1964, Bahnke and Howard [2] relaxed the assumption of zero axial conduction during flow exposures that was made by Lambertson [40]; however, they assumed negligible seal width in their analysis. They provided heat-transfer-performance design data in tabular form for regenerators with significant axial conduction but thin face seals. In the same year, a paper of London, et al. was published in which the authors correlated transient responses of regenerators to flow-temperature step changes.

A correlation of the effects of axial conduction during flow exposures by Shah [61] was published in 1975. His correlation accurately matches the data of Bahnke and Howard [2]. Such correlations are necessary for cycle calculations because high-accuracy calculation of effectiveness data using finite-difference integration is computationally intensive.

2.2.7 Flow Non-Uniformity

Header Design

Wilson [72] addressed flow non-uniformity due to non-ideal headers in 1966. He showed that reductions in regenerator effectiveness can be great for non-uniform flow distributions. He presented a method of design for headers. Using Wilson's procedure, the header designer can minimize flow non-uniformity through a specified core. London, et al. [44] also presented a header-design method. This method was published in 1968. London, et al. [44] derived shapes for inlet and outlet headers that produce uniform flow distributions.

Passage Uniformity

In 1970, London [43] addressed flow non-uniformity due to core manufacturing tolerances. He showed that this kind of flow non-uniformity can produce drastic reductions in effectiveness. Results such as his motivated new core manufacturing techniques (such as extrusion) that produce tight tolerances.

Modern regenerator cores typically are extruded with highly uniform regular passage geometries (see Day [19]). Therefore, theoretical results for regular passage geometries are applicable. The book Advances in Heat Transfer, "Laminar Flow Forced Convection in Ducts" was published in 1978 [63]. In this book, Shah and London summarize theoretical results for regular passage geometries and boundary conditions that obtain in regenerators (and other boundary conditions). Some work cited in the book of Shah and London is applicable to analysis (which is given in this book) of the effects of low porosity on heat-transfer performance [17, 48, 65, 69].

2.3 State of the Art

This chapter has described the development of regenerators and regenerator analysis and design. Today, industry is using the developed theoretical tools to design high-efficiency regenerative gas turbines (see Sayama [60]). The authors hope that the information given in this book will lead to optimized regenerative gas turbines for future automobiles, aircraft, marine propulsion, power generation, and other applications.

Chapter 3

Gas-Turbine Cycles

This book focuses on the design of regenerators for high-performance regenerative gas turbines. The ways in which gas-turbine regenerators can be designed for high system performance can be understood by studying gas-turbine cycles, the thermodynamic cycles by which gas turbines convert heat into work. This chapter introduces gas-turbine cycles.

Section 3.1 describes how to calculate two parameters that quantify the performance of a gas turbine: (1) thermal efficiency, η_{TH} and (2) cycle specific power, W'. Section 3.1 defines these performance parameters. The first and second laws of thermodynamics govern the performances of gas-turbine cycles, and determine the magnitudes of the two performance parameters. Section 3.2 gives expressions of the first and second laws.

The remainder of this chapter describes four kinds of gas-turbine cycles:

1. Simple cycle (Section 3.3);

2. Regenerative cycle (Section 3.4);

3. Intercooled Regenerative (ICR) cycle (Section 3.5); and

4. ICR cycle with reheat (Section 3.6).

The performances of these cycles are demonstrated with results from cycle calculations using the commercial software RGT-OPT™ [1]. These calculations show how gas turbines achieve high performance under the constraints imposed by the first and second laws of thermodynamics.

3.1 Performance

Thermal efficiency and cycle specific power are two parameters that quantify the performance of gas-turbine cycles.

[1]RGT-OPT is a trademark of Douglas Stephen Beck, the principal author, and is available from him.

Thermal Efficiency. Thermal efficiency indicates the amount of power produced by a gas turbine from a given heat-input rate:

$$\eta_{TH} \equiv \frac{\dot{W}}{\dot{Q}} , \tag{3.1}$$

where

η_{TH}	=	Thermal efficiency (–);
$\dot{W}$	=	Net power output of the gas turbine (W); and
$\dot{Q}$	=	Heat-input rate to the gas turbine (W).

From the first law of thermodynamics, the power output can at most equal the heat-input rate, so thermal efficiency can at most equal one. The second law of thermodynamics further constrains thermal efficiency. From the second law, Carnot [11] proved that the thermal efficiency of a gas-turbine cycle is limited as follows[2]:

$$\eta_{TH} \leq 1 - \frac{T_0}{TIT} , \tag{3.2}$$

where

T_0	=	Compressor-inlet temperature (K); and
TIT	=	Turbine-Inlet Temperature (K).

The upper limit on the thermal efficiency of a gas-turbine cycle is known as the Carnot efficiency:

$$\eta_{Carnot} \equiv 1 - \frac{T_0}{TIT} . \tag{3.3}$$

For example, for a compressor-inlet temperature of $T_0 = 300K$ and a turbine-inlet temperature of $TIT = 1,200K$, the Carnot efficiency is

$$\eta_{Carnot} = 1 - \frac{300K}{1,200K} = 75\% , \tag{3.4}$$

which is quite high. The efficiency of an actual engine is much lower than its associated Carnot efficiency. Efficiency goals for automotive regenerative gas turbines at full power are about 50%. Automotive spark-ignition engines have efficiencies of about half this level, and automotive diesel engines operate at about 30% efficiency.

Cycle Specific Power. In addition to thermal efficiency, cycle specific power is another parameter that quantifies the performance of gas-turbine cycles. Cycle specific power is calculated from the specific power of the components that comprise the gas turbine. The specific power of a component is defined as the

[2] Equation 3.2 is valid only if the temperatures are absolute temperatures. This book uses only absolute temperatures with Kelvin units.

power output of the component, divided by (roughly[3]) the enthalpy-flow rate into the compressor:

$$W'_j \equiv \frac{\dot{W}_j}{(\dot{m}c_P T)_0} , \tag{3.5}$$

where

W'_j	=	Component specific power (–);
$\dot{W}_j$	=	Power output of the component (W);
$\dot{m}$	=	Mass-flow rate (kg/s); and
c_P	=	Specific heat capacity at constant pressure ($J/kg-K$).

Cycle specific power is the sum of the specific powers of all n components that comprise the gas turbine:

$$W' \equiv \sum_{j=1}^{n} W'_j . \tag{3.6}$$

Therefore, cycle specific power is the net power output of a gas turbine, divided by (roughly) the enthalpy-flow rate into the compressor. Cycle specific power is a useful parameter for comparison of the performances of different engines. If two engines ingest identical flows of the same kind of gas at the same temperature and pressure, then the engine with the larger cycle specific power produces more power. If two engines produce the same power output, then the engine with the larger cycle specific power ingests less gas and is smaller.

A high-performance gas turbine has high thermal efficiency and high cycle specific power. However, the relative importance of each performance parameter depends on the particular application. Consider two different applications: (1) a rotorcraft gas turbine and (2) an automotive gas turbine. For rotorcraft gas turbines, high cycle specific power is nearly as important as high thermal efficiency. The goal when designing rotorcraft gas turbines is to minimize the combined weight of the engine and the fuel for a specified mission. Low fuel weight is achieved with engines with high thermal efficiencies and low fuel consumption. High specific powers are associated with small light-weight engines.

For automotive applications, high thermal efficiency is more important than high cycle specific power. A high engine thermal efficiency leads to a high automobile gas mileage. The size and weight of the engine is of less importance, because the size and weight of an automotive gas turbine typically are small, and the thermal efficiency of a small gas turbine increases as its size increases due to reductions in blade-tip-clearance losses and other losses.

This section has defined two parameters that are used to quantify the performance of gas-turbine cycles: (1) thermal efficiency and (2) cycle specific power. The next section (Section 3.2) gives the governing equations that are used to calculate these performance parameters.

---

[3] The denominator in Equation 3.5 is not precisely the enthalpy-flow rate into the compressor, because specific heat capacity varies with temperature for an ideal gas (see Section 3.2).

3.2 Governing Equations

The thermal efficiency and the cycle specific power of a gas turbine can be calculated using expressions of the first and second laws of thermodynamics. The expression of the first law is the Steady-Flow Energy Equation (SFEE). The expression of the second law is known as Gibbs' equation.

Steady-Flow Energy Equation. The Steady-Flow Energy Equation (SFEE) is an expression of the first law of thermodynamics that governs the steady-state operation of gas turbines (see Wilson [73] for a derivation of the SFEE[4]):

$$\dot{Q} - \dot{W} = \sum(\dot{m}i)_{out} - \sum(\dot{m}i)_{in} \ , \tag{3.7}$$

where

$\dot{Q}$	=	Net heat-transfer rate *into* a control volume (W);
$\dot{W}$	=	Net work-transfer rate *out of* the control volume (W);
$\dot{m}$	=	Mass-flow rate (kg/s);
i	=	Specific enthalpy (J/kg);
$\sum(\dot{m}i)_{out}$	=	Sum total of total-enthalpy-flow rates out of the control volume (W); and
$\sum(\dot{m}i)_{in}$	=	Sum total of total-enthalpy-flow rates into the control volume (W).

Most gas turbines (including all of the gas turbines considered in this book) use a working fluid that can be modelled as an ideal gas. For an ideal gas[5]

$$Pv = RT \ , \tag{3.8}$$

where

P	=	Pressure (Pa);
v	=	Specific volume (m^3/kg);
R	=	Gas constant for the particular gas ($J/kg - K$); and
T	=	Absolute temperature (K).

Also for an ideal gas[6],

$$di = c_P dT \ , \tag{3.9}$$

and

$$du = c_V dT \ , \tag{3.10}$$

[4]The difference between total properties and static properties in Equation 3.7 is that total properties include effects of the kinetic energy (see Wilson [73] for a more rigorous definition and explanation of this difference.) We use total properties throughout this book unless we denote the properties with the subscript "st" for static.

[5]Note in Equation 3.8 that $R \equiv R_U/MW$, where R_U ($J/kgmol - K$) is the universal gas constant and MW ($kg/kgmol$) is the molecular weight of the gas.

[6]Note that $c_V = c_P - R$.

where

i = Specific enthalpy (J/kg);
c_P = Specific heat capacity at constant pressure $(J/kg-K)$;
T = Temperature (K);
u = Specific internal energy (J/kg); and
c_V = Specific heat capacity at constant volume $(J/kg-K)$.

Chappell and Cockshutt [12] give data for the specific heat capacities and for the specific enthalpies of dry air and of combustion products as functions of temperature. Correlations of these data can be used to accurately predict gas-turbine performance. (The commercial software RGT-OPT uses correlations of these data, and RGT-OPT was used throughout this book for cycle calculations and regenerator optimizations.)

Gibbs' Equation. The SFEE (Equation 3.7) is an expression of the first law of thermodynamics. The following is Gibbs' equation, which is an expression of the second law of thermodynamics:

$$Tds = du + Pdv \ , \tag{3.11}$$

where

T = Temperature (K);
s = Specific entropy $(J/kg-K)$;
u = Specific internal energy (J/kg);
P = Pressure (Pa); and
v = Specific volume (m^3/kg).

Equations 3.8–3.10 are used in conjunction with Equation 3.11 to yield an explicit expression for specific entropy:

$$s = s_{REF} + \int_{T_{REF}}^{T} \frac{c_P(\sigma)}{\sigma} d\sigma - R \ln \left(\frac{P}{P_{REF}} \right) \ , \tag{3.12}$$

where

s_{REF} = Reference specific entropy $(J/kg-K)$ at a reference temperature, T_{REF} (K), and a reference pressure, P_{REF} (Pa).

The selection of a reference state is arbitrary; however, specific entropy is usually set equal to zero at atmospheric temperature and pressure. Thus, the air that enters a compressor usually is assigned a specific entropy of zero.

The expression for specific entropy given by Equation 3.12 can be used to calculate the outlet conditions of turbo-machines with known efficiencies. It can also be used to generate temperature-vs.-entropy plots, which are used in the

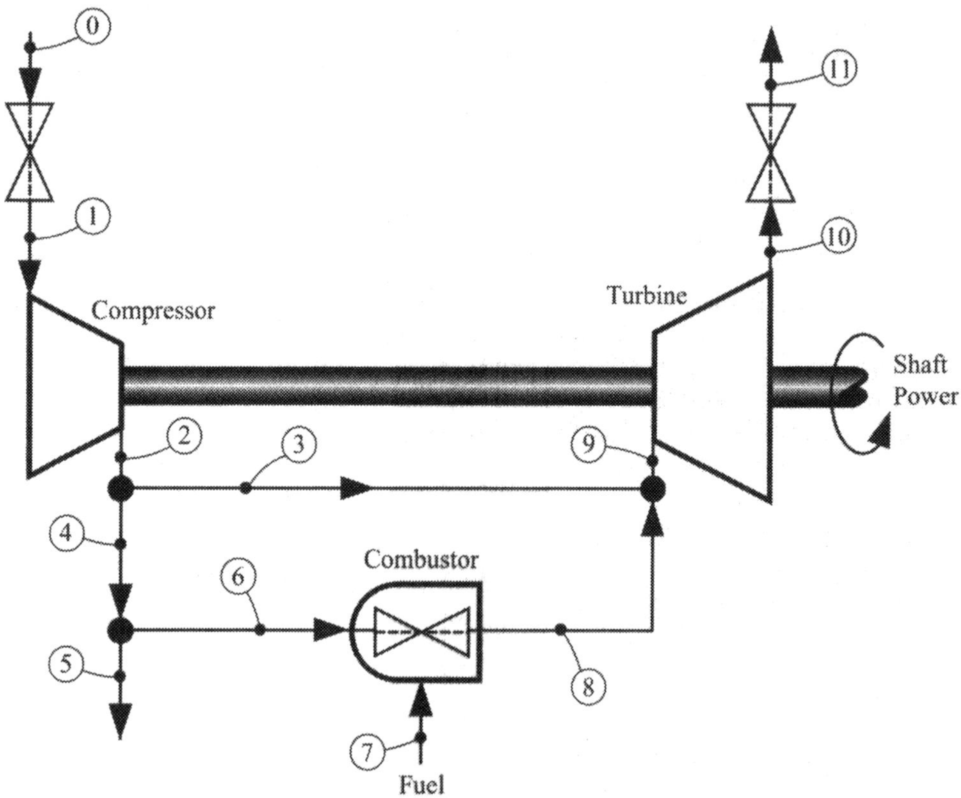

Figure 3.1: Simple-Cycle Gas Turbine.

rest of the sections in this chapter to illustrate the performances of gas-turbine cycles.

The rest of this chapter shows how gas-turbine cycles achieve high performances (quantified by large thermal efficiencies and specific powers) under the constraints imposed by the first and second laws of thermodynamics.

3.3 Simple Cycle

Figure 3.1 is a schematic of a simple-cycle gas turbine. Various points in the cycle are numbered in Figure 3.1. A simple-cycle gas turbine is comprised of a turbo-compressor, a combustor, and a turbo-expander (or turbine).

Consider the flow of gas through a simple-cycle gas turbine. Air enters the engine at atmospheric temperature and pressure. The compressor compresses the air to a high pressure. The air then flows through the combustor where fuel is added and burned to achieve a high temperature at the inlet to the turbine. The turbine extracts power from the hot gas as it expands from a high pressure down to atmospheric pressure. The principle of operation of gas turbines is that the amount of power required to compress cool gas is less than the amount of power that can be extracted during an expansion at high temperatures. Thus, only some (not all) of the power generated by the turbine in a gas-turbine engine is used to drive the compressor. The rest is useful shaft power.

An actual gas turbine contains leakages and pressure drops that reduce the performance of the engine. Most actual simple-cycle engines can be modelled using the leakage paths and pressure drops indicated in Figure 3.1. (Figure 3.1 does not show *all* possible leakage paths and pressure drops, but most losses can be modelled using the finite number of losses indicated.)

Pressure Drops. Pressure drops in a gas turbine reduce engine performance because increased pumping power is required by the compressor to drive the flows through the resistances with the associated pressure drops. Consider pressure drops at the gas-turbine inlet. With the simple-cycle gas turbine shown in Figure 3.1, air enters the engine at Point 0. The air may then flow through one or more pre-conditioning devices with associated pressure drops from Point 0 to Point 1 in Figure 3.1. These devices include:

1. Compressor inlet headers;
2. Air filters for removing dirt particles, which could damage the compressor;
3. Pre-coolers for cooling the inlet air so less work is required by the compressor to compress the cooler air; and
4. Pre-heaters for heating the inlet air to prevent ice formation on the compressor's inlet guide vanes (at the expense of an increase in the power required by the compressor).

The pressure drop associated with these devices (as well as other pressure drops in a gas turbine) can be quantified by a dimensionless pressure drop, which is often referred to as "Delta P on P":

$$\frac{\Delta P}{P} \equiv \frac{(P_I - P_O)}{0.5(P_I + P_O)}, \tag{3.13}$$

where

$\frac{\Delta P}{P}$ = Dimensionless pressure drop (–);
P_I = Inlet pressure (Pa); and
P_O = Outlet pressure (Pa).

Figure 3.2 shows a temperature-vs.-entropy plot of a simple cycle with the component specifications listed in Table 3.1. The numbered points in Figure 3.2 correspond with the numbered points in the engine schematic in Figure 3.1. During the pressure drop from Point 0 to Point 1 the total temperature of the inlet air is constant, but the pressure decreases and the specific entropy increases. Table 3.1 shows that the inlet pressure drop is 2%, which is too small to be detected in Figure 3.2.

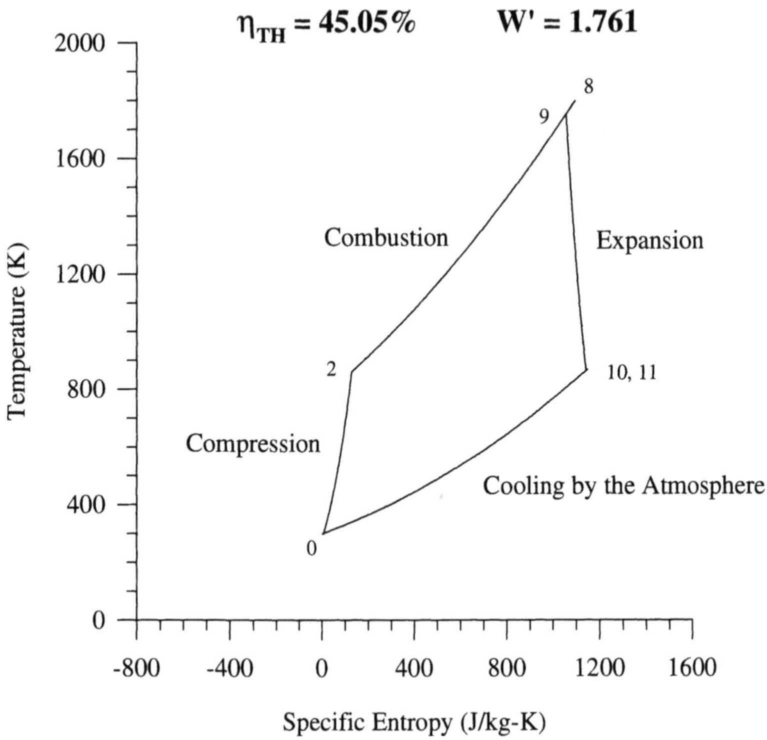

Figure 3.2: Design-Point Performance and Plot of Temperature vs. Specific Entropy for a Simple-Cycle Gas Turbine with Component Specifications Listed in Table 3.1 (data calculated by RGT-OPT).

Component Specification	Symbol	Value
Turbine Inlet Temperature	TIT	$1,800K$
Compressor Pressure Ratio	r_C	30
Compressor Polytropic Total-to-Total Efficiency	η_{p00C}	89%
Turbine Polytropic Total-to-Total Efficiency	η_{p00E}	91%
Dimensionless Combustor Pressure Drop	$(\Delta P/P)_H$	4%
Turbine-Blade-Cooling Constant	$K_{Cooling}$	5%
Turbine Blade Temperature	T_{Blade}	$1,300K$
Dimensionless Inlet Pressure Drop	$(\Delta P/P)_I$	2%
Dimensionless Outlet Pressure Drop	$(\Delta P/P)_O$	2%
Dimensionless Leakage	$\dot{m}_{Leak}/\dot{m}_4$	1%
Atmospheric Temperature	T_{atm}	300 K

Table 3.1: Component Specifications for the Simple Cycle Gas Turbine Shown in Figure 3.1.

Compression. From Point 1 to Point 2 in Figures 3.1 and 3.2, the compressor compresses the air to a high pressure, and (therefore) the temperature and entropy of the air increase. The temperature at Point 2 can be calculated from the polytropic total-to-total efficiency of the compressor:

$$T_2 = T_1 \left(r_C^{\frac{R}{\bar{c}_P \eta_{p00C}}} \right) , \tag{3.14}$$

where

T_2	=	Temperature at the compressor outlet (K);
T_1	=	Temperature at the compressor inlet (K);
r_C	=	Compressor pressure ratio (–), the ratio of the outlet pressure, P_2 (Pa), to the inlet pressure, P_1 (Pa);
R	=	Gas constant for air $(J/kg - K)$;
$\bar{c}_P$	=	Average specific heat capacity at constant pressure for air $(J/kg - K)$; and
η_{p00C}	=	Polytropic total-to-total efficiency of the compressor (–).

Polytropic efficiency is defined as the isentropic efficiency of a compression process through a pressure ratio approaching one. Thus, polytropic efficiency is sometimes referred to as "Small-stage efficiency", and it can be thought of as the isentropic efficiency of each of many small compression processes in succession that form a complete compression process. "Total-to-total" in the efficiency "Polytropic total-to-total efficiency" indicates that total conditions at the inlet and outlet flange of the compressor are considered in the efficiency definition. A polytropic total-to-total efficiency of one is associated with an isentropic (constant entropy) compression, which requires the least power. Compressors with lower efficiencies require more power to achieve a given pressure ratio. Typical efficiencies are:

1. η_{p00C} = 83% for small radial-flow compressors with pressure ratios of 3 (for example the compressors used in automotive gas turbines); and

2. η_{p00C} = 89% for large axial-flow compressors with pressure ratios of 30 (for example the compressors of aircraft gas turbines).

Figure 3.2 shows how the temperature and entropy of the air increase during the actual compression process. An ideal isentropic (constant entropy) compression would have a smaller temperature increase. By combining the Steady-Flow Energy Equation (SFEE) and the relation for the specific enthalpy of an ideal gas (Equations 3.7 and 3.9) we can show that the isentropic compression (with a polytropic total-to-total efficiency of one) would require less work than actual compressions with lower efficiencies:

$$\dot{Q} - \dot{W} = \dot{m}\bar{c}_P \left(T_{out} - T_{in} \right) , \tag{3.15}$$

where

$\dot{Q}$	=	Net heat-transfer rate *into* the component—usually negligible for compressors (W);
$\dot{W}$	=	Net work-transfer rate *out of* the component—negative for compressors, which absorb work (W);
$\dot{m}$	=	Mass-flow rate through the component (kg/s);
$\bar{c}_P$	=	Average specific heat capacity of the gas flowing through the component ($J/kg - K$);
T_{out}	=	Total temperature of the gas flowing out of the component (K); and
T_{in}	=	Total temperature of the gas flowing into the component (K).

Equation 3.15 shows that the power absorbed by a compressor is proportional to the temperature rise of the air from inlet to outlet. An isentropic compressor would have the smallest temperature rise, so it would require the smallest power input.

Equation 3.15 is strictly valid only if the average specific heat capacity is calculated as follows:

$$\bar{c}_P = \frac{\int_{T_{in}}^{T_{out}} c_P(\sigma)d\sigma}{T_{out} - T_{in}} . \tag{3.16}$$

However, calculations are sufficiently accurate if the average specific heat capacity is calculated from the mean temperature in the component:

$$\bar{c}_P \approx c_P\left(\frac{T_{out} + T_{in}}{2}\right) . \tag{3.17}$$

Turbine Cooling. The Carnot efficiency of a gas turbine is high for hot turbine-inlet temperatures, TIT (see Equation 3.3). Actual thermal efficiencies are also high. In addition, gas turbines with hot turbine-inlet temperatures have high specific powers. This has led manufacturers of large gas turbines to use turbine-inlet temperatures that are significantly hotter than the maximum temperatures that the turbine blades can withstand and still have acceptable life. Turbine-inlet temperatures for aircraft gas turbines (for which high specific power and high efficiency are more important than long life) are typically about $TIT \approx 1,700$–$1,800K$. Turbine-inlet temperatures for power-generation gas turbines (for which long life is more important) are typically about $TIT \approx 1,400$–$1,500K$. The maximum temperature to which a metallic blade can be heated and still have an acceptable life is $T_{Blade} \approx 1,300K$. The engines can achieve hot turbine-inlet temperature *and* have acceptable engine life if the turbine blades and nozzles are cooled.

For large gas turbines ($> 1MW$), a small amount of compressed air is bled from the compressor discharge to cool the turbine blades. Figure 3.1 shows a turbine-cooling flow (Point 3) being bled from the compressor discharge (Point 2). Small gas turbines like automotive gas turbines ($\approx 75kW$), must be cost-competitive with spark-ignition and diesel engines. It would be too expensive to cool the turbine blades of small gas turbines. Therefore, small gas turbines use ceramic turbines rather than turbine-blade cooling to achieve hot turbine-inlet temperatures ($TIT \approx 1400K$).

For a gas turbine to have high performance, the amount of compressed air bled to cool the turbine blades must be kept to a minimum. Compressed air that is bled to the turbine does not pass through the combustor, so only part of the power required to compress this gas in the compressor is recovered by the turbine. Therefore, the small amount of compressed air bled from the compressor discharge (typcially less than 10% of the main flow) must be highly effective in cooling the turbine blades. A blade-cooling effectiveness is defined as follows[7]:

$$\epsilon_{BC} \equiv \frac{TIT - T_{Blade}}{TIT - T_2} , \tag{3.18}$$

where

ϵ_{BC} = Blade-cooling effectiveness (–);
TIT = Turbine-inlet temperature (K);
T_{Blade} = Maximum blade temperature (K); and
T_2 = Compressor discharge temperature (K).

For a perfect blade-cooling effectiveness of one, the coolant would keep the turbine blades at the compressor-discharge temperature. If the blade-cooling effectiveness were zero, then the coolant would not cool the blades at all, and the temperature of the turbine blades would be equal to the turbine-inlet temperature (TIT).

A simple analysis shows how blade-cooling effectiveness is related to the required amount of coolant. From conservation of energy, the amount of heat absorbed by the coolant as it flows through the turbine blade is equal to the heating of the blade by the main flow. For this simple analysis, we can assume that the coolant exits the turbine blade at the blade temperature, T_{Blade}. The heating of the blade can be calculated from Newton's law of convective heat transfer. Therefore,

$$\dot{m}_{Coolant}\bar{c}_P \left(T_{Blade} - T_2\right) = (hA)_{Surface} \left(TIT - T_{Blade}\right) , \tag{3.19}$$

where

$\dot{m}_{Coolant}$ = Coolant mass-flow rate (kg/s);
$\bar{c}_P$ = An average specific heat capacity ($J/kg - K$); and
$(hA)_{Surface}$ = Convective conductance associated with the blade surface exposed to the main gas flow (W/K).

The convective conductance associated with the blade surface, $(hA)_{Surface}$, is roughly proportional to the mass-flow rate of the main turbine gas flow,

[7] In Equation 3.18 the turbine-inlet temperature is used for the temperature of the gas stream flowing through the turbine. This approximation is accurate because most of the cooling occurs in the first few turbine stages where the gas temperature is approximately equal to the turbine-inlet temperature.

$\dot{m}_{Turbine}$. Therefore, the coolant requirement is related to the blade-cooling effectiveness as follows:

$$\frac{\dot{m}_{Coolant}}{\dot{m}_{Turbine}} = K_{Cooling}\left(\frac{\epsilon_{BC}}{1-\epsilon_{BC}}\right) , \tag{3.20}$$

where

$K_{Cooling}$ = A turbine-blade-cooling constant (–).

The turbine-blade-cooling constant, $K_{Cooling}$, depends on the cooling technology. Typical values are:

1. $K_{Cooling} = 3\%$ for state-of-the-art cooling technologies;
2. $K_{Cooling} = 5\%$ for standard cooling technologies; and
3. $K_{Cooling} = 7\%$ for relatively poor cooling technologies.

In an actual gas turbine, the coolant is injected into the main flow through the first few stages of nozzles and blades. We can model this by assuming that the coolant mixes with the main flow before the main flow enters the turbine. In this model, the coolant reduces the turbine-inlet temperature, which reduces thermal efficiency and specific power. (Note that this reduced turbine-inlet temperature is still much higher than if the blades were not cooled at all.) The engine schematic in Figure 3.1 shows this model: coolant (Point 3) mixes with the hot flow from the combustor (Point 8) to produce a turbine-inlet flow (Point 9) of reduced temperature.

Figure 3.2 shows the effect that turbine cooling has on the performance of gas-turbine cycles. The coolant reduces the temperatures in the turbine, so the power output of the turbine is reduced.

Leakage. Actual gas turbines have small leakages that reduce engine performance. Figure 3.1 shows a leakage (Point 5) from the compressor discharge (Point 4) to the ambient environment. This is a typical leakage. The effect of any leakage is to reduce the power output of the turbine. Any leakage represents power absorbed by the compressor without power generated by the turbine. Figure 3.2 is a plot of temperature vs. *specific* entropy (entropy per unit mass), so the effect of leakage is not shown. However, as Equation 3.15 shows, for a fixed temperature rise across the compressor and a fixed temperature drop across the turbine, the power input to the compressor will stay the same but the turbine's power output will drop if the flow through the compressor is maintained and the flow through the turbine is reduced.

Combustion. Most gas turbines employ internal combustion to raise the temperature of the compressed air to a high turbine-inlet temperature (TIT). A higher turbine-inlet temperature is possible with internal combustion than when a heat exchanger is used to transfer heat into the gas turbine (external combustion). Usually the fuel flow is only a small percentage of the main gas flow with internal combustion (fuel-to-air ratios of about 1–4%), because a large amount of excess air is required to maintain the turbine inlet temperature at an acceptably low value. Compared with external combustion, internal combustion

1. Increases the flow through the turbine, which tends to increase the turbine's power output; and
2. Increases the specific heat capacity of the gas, which tends to reduce the turbine's power output.

These two effects essentially cancel each other to produce no significant improvement in engine performance. The correlations of Chappell and Cockshutt [12] can be used to calculate the amount of fuel that must be added to the combustor to achieve a given turbine-inlet temperature, as well as the properties of the resulting products of combustion.

Combustors must reduce the speed of the flow delivered by the compressor from about 100–200 m/s down to the flame speed associated with the combustion, which is about 10 m/s (see Wilson [72]). This requires a significant drop in total pressure. A typical value is $\Delta P/P \approx 4\%$. Figure 3.2 shows heat addition via internal combustion from Point 6 (which has the same temperature and specific entropy as Point 2) to Point 8.

Expansion. From Point 9 to Point 10 in Figures 3.1 and 3.2, the turbine extracts power from the hot high-pressure gas that flows out of the combustor. As the gas expands to near atomspheric pressure, its temperature decreases and its specific entropy increases. The outlet temperature (at Point 10) can be calculated similar to the way we calculate the outlet temperature of a compressor:

$$T_{10} = T_9 \left(r_E^{\frac{-R\eta_{p00E}}{\bar{c}_P}} \right), \tag{3.21}$$

where

T_{10}	=	Temperature at the turbine outlet (K);
T_9	=	Temperature at the turbine inlet (K);
r_E	=	Turbine pressure ratio (–), the ratio of the inlet pressure, P_9 (Pa), to the outlet pressure, P_{10} (Pa);
R	=	Gas constant for combustion products ($J/kg-K$);
$\bar{c}_P$	=	Specific heat capacity at constant pressure for combustion products ($J/kg-K$); and
η_{p00E}	=	Polytropic total-to-total efficiency of the turbine (–).

The polytropic total-to-total efficiency for a turbine is defined in a similar way to that for a compressor. Typical efficiencies are:

1. $\eta_{p00E} = 89\%$ for small radial-in-flow turbines with pressure ratios of 3 (for example the turbines used in some automotive gas turbines); and

2. $\eta_{p00E} = 91\%$ for large axial-flow turbines with pressure ratios of 30 (for example the turbines of aircraft gas turbines).

Exit Pressure Drop. Gas turbines often have components downstream of the turbine that provide back pressure on the turbine (that increase the pressure at the turbine exit to above atmospheric pressure). These components include:

1. Turbine exit headers;

2. Heat-Recovery Boilers (HRB's) or Heat-Recovery Steam Generators (HRSG's) in combined cycles; and

3. Pollution-control devices (such as scrubbers).

By increasing the pressure at the turbine exit, these components reduce the expansion ratio of the turbine, which reduces the power output of the turbine. However, these components are often unavoidable (for example pollution-control devices), and HRSG's are used in combined cycles, which are currently the most efficient heat engines in the world with thermal efficiencies approaching 60%.

The simple-cycle gas turbine in Figure 3.1 has an exit pressure drop from Point 10 to Point 11. The entropy increases but total temperature is constant as the gas flows through the exit pressure drop. The changes are too small to be detected in Figure 3.2.

Simple Cycle with a Low Pressure Ratio. A simple-cycle gas turbine with a high pressure ratio and a high turbine-inlet temperature can achieve a high thermal efficiency and a high specific power. (Note that the cycle in Figure 3.1 has a thermal efficiency of $\eta_{TH} = 45.05\%$ and a specific power of $W' = 1.761$.) Large ($\approx 60MW$) aircraft gas turbines can achieve high pressure ratios with multiple compressor stages. Also, complex turbine-blade-cooling technologies can be employed to permit turbine-inlet temperatures that are well above the maximum temperatures that the metal turbine nozzles and blades can withstand and still have acceptable life.

In small ($< 200kW$) gas turbines it is difficult to achieve high pressure ratios efficiently because the compressor blades in the high-pressure portions of the compressor are small and blade-tip leakages are large. In addition, small components and cost considerations preclude turbine cooling, so turbine-inlet temperatures must be kept lower in small gas turbines than in large gas turbines. This is likely to continue to be true even when nonmetals, such as ceramics, are used for turbines and casings.

Automotive gas turbines typically have pressure ratios less than five and turbine-inlet temperatures of about $1,400K$ for ceramic turbines. Table 3.2 lists typical component specifications. Figure 3.3 shows that a simple-cycle automotive gas turbine would have an efficiency of only 22.98% at design point. The efficiency at part-power would be even worse. There is no way that such an engine could compete with diesel or spark-ignition engines.

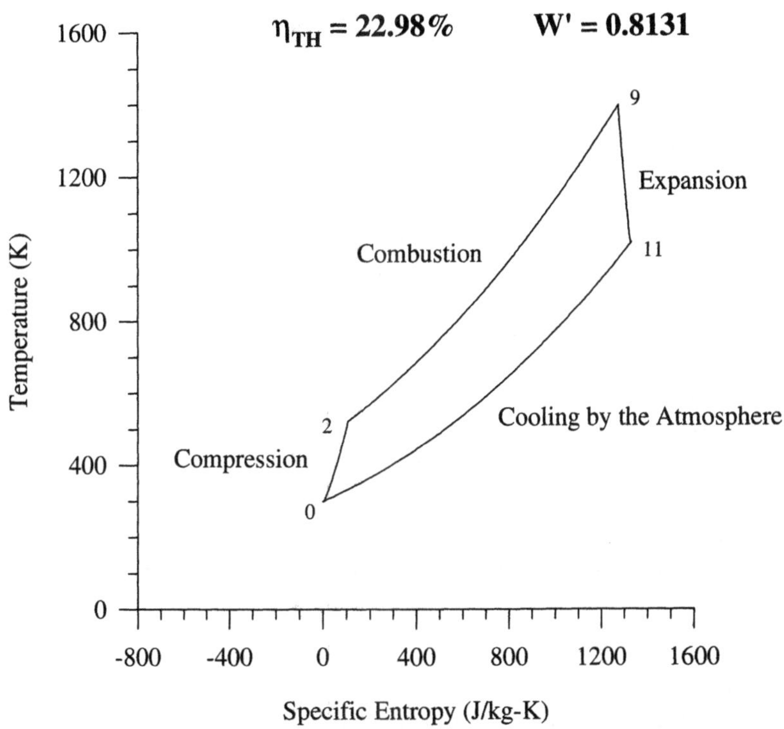

Figure 3.3: Design-Point Performance and Plot of Temperature vs. Specific Entropy for a Low-Pressure-Ratio Simple-Cycle Gas Turbine with Component Specifications Listed in Table 3.2 (data calculated by RGT-OPT).

Component Specification	**Symbol**	**Value**
Turbine Inlet Temperature	TIT	$1,400K$
Compressor Pressure Ratio	r_C	30
Compressor Polytropic Total-to-Total Efficiency	η_{p00C}	82%
Turbine Polytropic Total-to-Total Efficiency	η_{p00E}	88%
Dimensionless Combustor Pressure Drop	$(\Delta P/P)_H$	4%
Turbine-Blade-Cooling Constant	$K_{Cooling}$	0%
Turbine Blade Temperature	T_{Blade}	$1,400K$
Dimensionless Inlet Pressure Drop	$(\Delta P/P)_I$	2%
Dimensionless Outlet Pressure Drop	$(\Delta P/P)_O$	2%
Dimensionless Leakage	$\dot{m}_{Leak}/\dot{m}_4$	1%
Atmospheric Temperature	T_{atm}	300 K

Table 3.2: Component Specifications for a Low-Pressure-Ratio Simple-Cycle Gas Turbine.

Component Specification	Symbol	Value
Turbine Inlet Temperature	TIT	$1,400K$
Compressor Pressure Ratio	r_C	5
Compressor Polytropic Total-to-Total Efficiency	η_{p00C}	82%
Turbine Polytropic Total-to-Total Efficiency	η_{p00E}	88%
Dimensionless Combustor Pressure Drop	$(\Delta P/P)_H$	4%
Turbine-Blade-Cooling Constant	$K_{Cooling}$	0
Turbine Blade Temperature	T_{Blade}	$1,400K$
Dimensionless Inlet Pressure Drop	$(\Delta P/P)_I$	2%
Dimensionless Outlet Pressure Drop	$(\Delta P/P)_O$	2%
Dimensionless Regenerator Inlet Leakage	$\dot{m}_{XLI}/\dot{m}_4$	2%
Dimensionless Regenerator Outlet Leakage	$\dot{m}_{XLO}/\dot{m}_{13}$	2%
Atmospheric Temperature	T_{atm}	300 K
Regenerator Effectiveness	ϵ_X	95%
Dimensionless Pressure Drop Across the Compressed-Air Side of the Regenerator	$(\Delta P/P)_N$	2%
Dimensionless Pressure Drop Across the Exhaust Side of the Regenerator	$(\Delta P/P)_X$	2%

Table 3.3: Component Specifications for the Regenerative Gas Turbine Shown in Figure 3.4.

Fortunately, it is possible to include a heat exchanger in gas turbines to recover heat from the exhaust and achieve a high thermal efficiency. This gas-turbine cycle is known as the regenerative cycle. The next section describes how regenerative cycles achieve high performance.

3.4 Regenerative Cycle

Adding a regenerator to a gas turbine can increase the engine's performance. Other chapters of this book describe regeneration in detail. This section illustrates the effects of regeneration on the performance of gas turbines.

Figure 3.4 shows a regenerative gas turbine, and Figure 3.5 gives its performance. The regenerator recovers heat in the exhaust (from Point 15 to Point 16) and uses this thermal energy to preheat the compressed air (from Point 12 to Point 13) before it enters the combustor (at Point 6). Thus, thermal efficiency increases from 23% for a low-pressure-ratio simple cycle to 43% for an otherwise-identical cycle with a 95%-effective regenerator (compare Figures 3.3 and 3.5).

An ideal heat exchanger (with ideal performance—an effectiveness of $\epsilon = 1$, no leakage, no pressure drops, and no power consumption) would raise the temperature of the compressed air up to the temperature at the turbine outlet. Actual regenerators have losses that limit their performance. An actual heat exchanger transfers a smaller amount of heat to the compressed air than an ideal

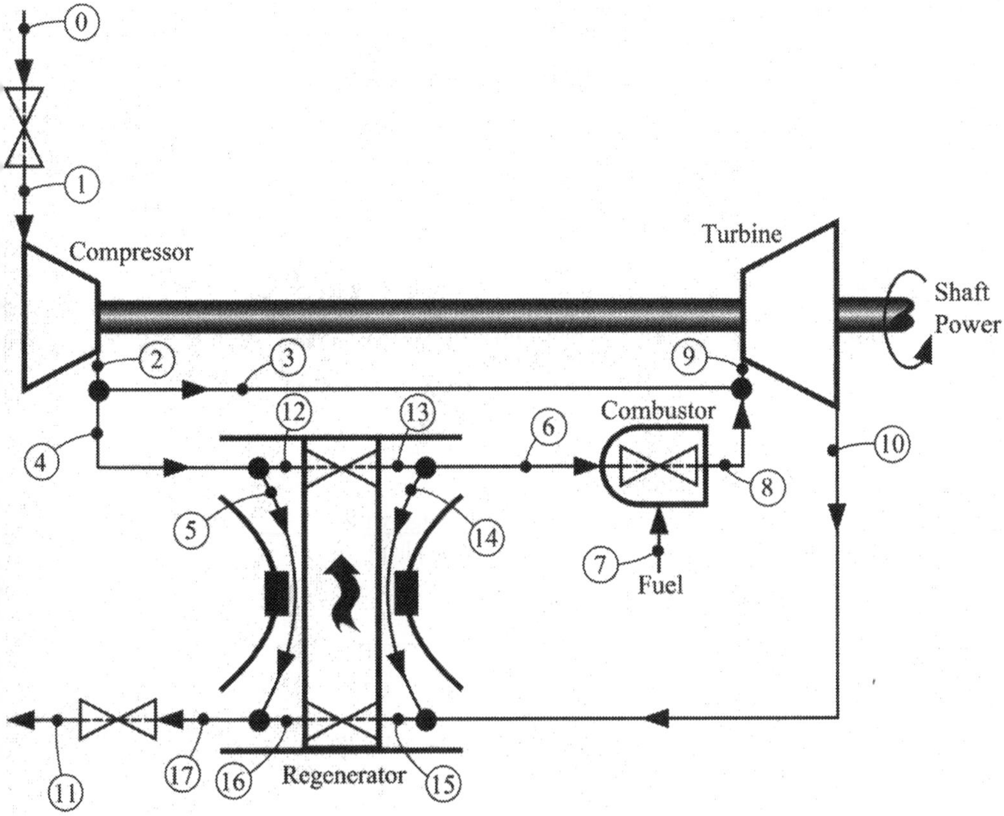

Figure 3.4: Regenerative Gas Turbine.

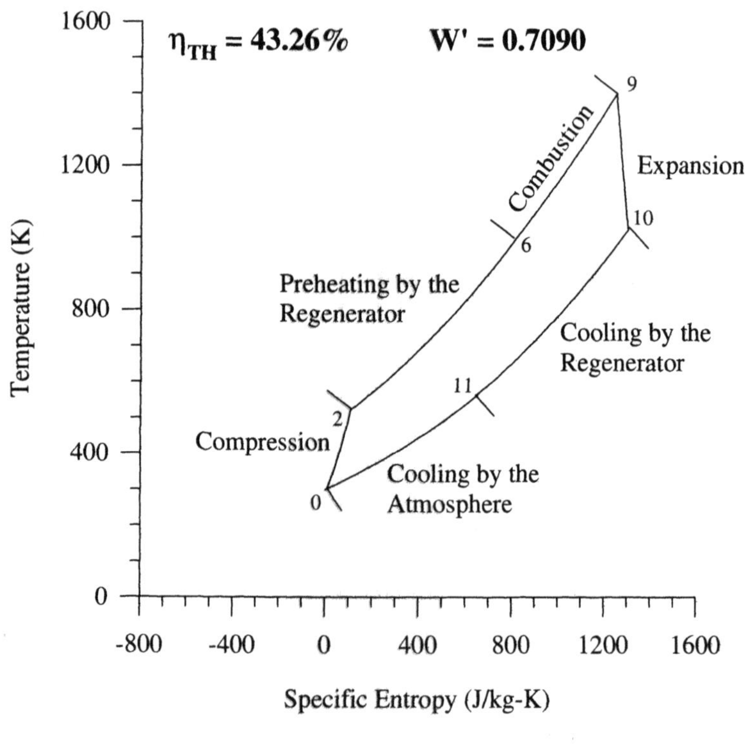

Figure 3.5: Design-Point Performance and Plot of Temperature vs. Specific Entropy for a Regenerative Gas Turbine with Component Specifications Listed in Table 3.3 (data calculated by RGT-OPT).

heat exchanger. Also, the compressed air and exhaust experience pressure drops as they flow across the heat-transfer surfaces on each side of the regenerator.

In addition to engine-performance penalties due to non-ideal heat transfer and pressure drops, regenerators add two other penalties to engine performance:

1. Compressed gas is lost through seal leakage; and

2. A regenerator requires a finite power input to rotate the core.

Because of regenerator seal leakage, the mass-flow rate through the compressor is greater than the mass-flow rate through the expander, and the specific power and the thermal efficiency of the cycle are reduced. Also, an actual regenerator requires a finite power input, which detracts from the power output of the engine and reduces thermal efficiency and cycle specific power.

The heat transfer, leakage, pressure drops, and power consumption of a regenerator conflict. For example a high-effectiveness regenerator requires large flow areas (see Chapter 5), but it is difficult to diffuse the flows from the high velocities at the compressor and turbine exits to the low velocities associated with these large flow areas. Therefore, the pressure drops associated with the header/core systems can be large for a high-effectiveness regenerator. In addition, a high-effectiveness regenerator can have large seal areas (and high leakage), and could require a significant power input to turn its large and heavy core. Maximum cycle performance is achieved as the result of a trade-off of heat transfer, leakages, pressure drops, and power consumption. This trade-off guides the design of gas-turbine regenerators, as described in Chapter 5.

Including a regenerator in a gas turbine improves engine performance. The next section shows that intercooling the compressor can lead to further gains.

3.5 Intercooled Regenerative (ICR) Cycle

Intercooling the compressor in a gas turbine reduces the amount of power absorbed by the compressor because less power is expended compressing cool gas than hot gas. Thus the power output and cycle specific power of the gas turbine increase. However, unless a regenerator is included in the cycle, thermal efficiency will be low because intercooling reduces the compressor-delivery temperature, so more fuel will need to be added in the combustor to achieve a given turbine-inlet temperature. A well known Intercooled Regenerative (ICR) gas turbine is the Westinghouse/Rolls-Royce WR-21 (see [70]).

Figure 3.6 is a schematic of an ICR gas turbine, and Figure 3.7 is a performance plot of this cycle. The compression process has been divided into two stages with an intercooler between the stages. An ideal intercoooler (with a perfect effectiveness of one) would reduce the temperature of the air from the first stage (Point 19) down to the temperature of the intercooler's heat sink (Point 25). An actual intercooler has an effectiveness less than one and only achieves partial cooling of the compressed air. In addition, the compressed air experiences a finite pressure drop as it flows across the heat-transfer surfaces

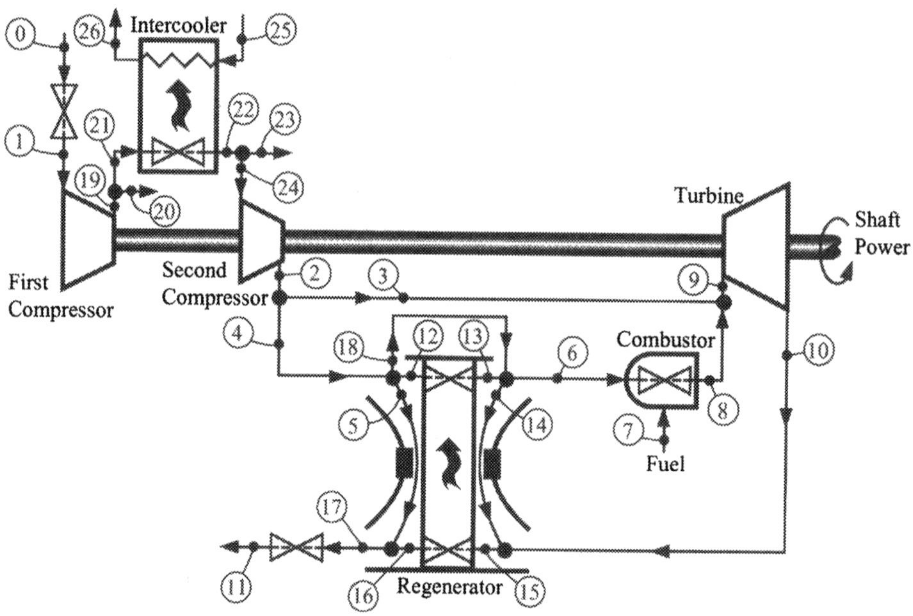

Figure 3.6: Intercooled Regenerative (ICR) Gas Turbine.

in an actual intercooler and some of the compressed air is lost to the ambient environment at Points 20 and 23. These losses reduce engine thermal efficiency and cycle specific power.

3.6 ICR Cycle with Reheat

As described above, the principle of operation of a gas turbine is that the amount of power expended compressing cool gas is less than the amount of power extracted during an expansion process at high temperatures. Thus the performance of a gas turbine can be increased by cooling the gas during the compression process and/or reheating the gas during the expansion process. We saw in the last section (Section 3.5) how cooling the gas during the compression process (intercooling) increases the performance of an Intercooled Regenerative (ICR) gas turbine. This section describes reheating during the expansion process. The Ford 704 gas turbine for trucks included a reheat combustor and an intercooler (see Wilson [83]).

Figure 3.8 shows an ICR gas turbine with a reheat combustor, and Figure 3.9 is a performance plot of this cycle. The first combustor heats the gas from Point 6 to a first turbine-inlet temperature, TIT_1, at Point 8. The gas cools from

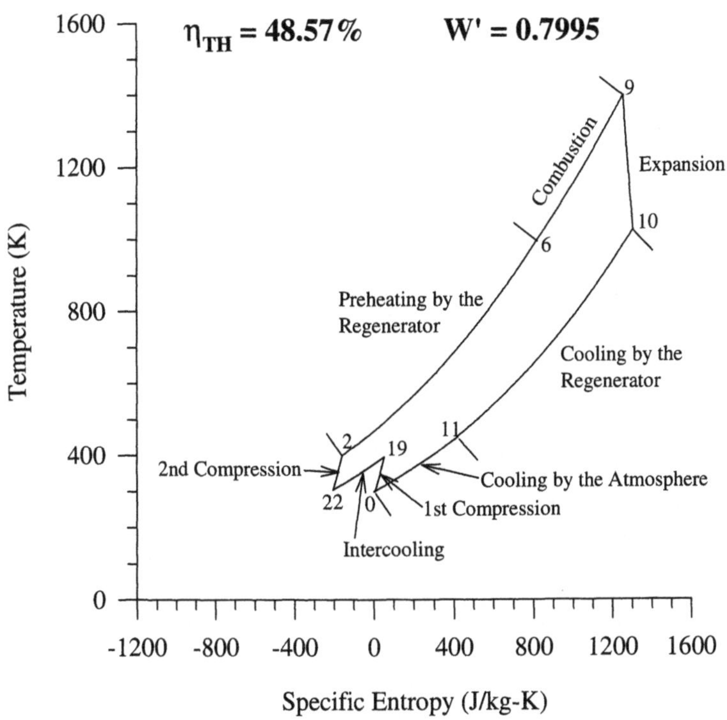

Figure 3.7: Design-Point Performance and Plot of Temperature vs. Specific Entropy for an ICR Gas Turbine with Component Specifications Listed in Table 3.4 (data calculated by RGT-OPT).

Component Specification	**Symbol**	**Value**
Turbine Inlet Temperature	TIT	$1,400K$
Pressure Ratio of the First Compressor Stage	r_{C1}	2.25
Polytropic Total-to-Total Efficiency of the First Compressor Stage	η_{p00C1}	83%
Turbine Polytropic Total-to-Total Efficiency	η_{p00E}	88%
Dimensionless Combustor Pressure Drop	$(\Delta P/P)_H$	4%
Turbine-Blade-Cooling Constant	$K_{Cooling}$	0
Turbine Blade Temperature	T_{Blade}	$1,400K$
Dimensionless Inlet Pressure Drop	$(\Delta P/P)_I$	2%
Dimensionless Outlet Pressure Drop	$(\Delta P/P)_O$	2%
Dimensionless Regenerator Inlet Leakage	$\dot{m}_{XLI}/\dot{m}_4$	2%
Dimensionless Regenerator Outlet Leakage	$\dot{m}_{XLO}/\dot{m}_{13}$	2%
Atmospheric Temperature	T_{atm}	300 K
Regenerator Effectiveness	ϵ_X	95%
Dimensionless Pressure Drop Across the Compressed-Air Side of the Regenerator	$(\Delta P/P)_N$	2%
Dimensionless Pressure Drop Across the Exhaust Side of the Regenerator	$(\Delta P/P)_X$	2%
Intercooler Effectiveness	ϵ_I	95%
Dimensionless Pressure Drop Across the Air Side of the Intercooler	$(\Delta P/P)_I$	2%
Dimensionless Intercooler Inlet Leakage	$\dot{m}_{ILI}/\dot{m}_{19}$	0
Dimensionless Intercooler Outlet Leakage	$\dot{m}_{ILO}/\dot{m}_{22}$	0
Pressure Ratio of the Second Compressor Stage	r_{C2}	2.22
Polytropic Total-to-Total Efficiency of the Second Compressor Stage	η_{p00C2}	83%

Table 3.4: Component Specifications for the ICR Gas Turbine Shown in Figure 3.6.

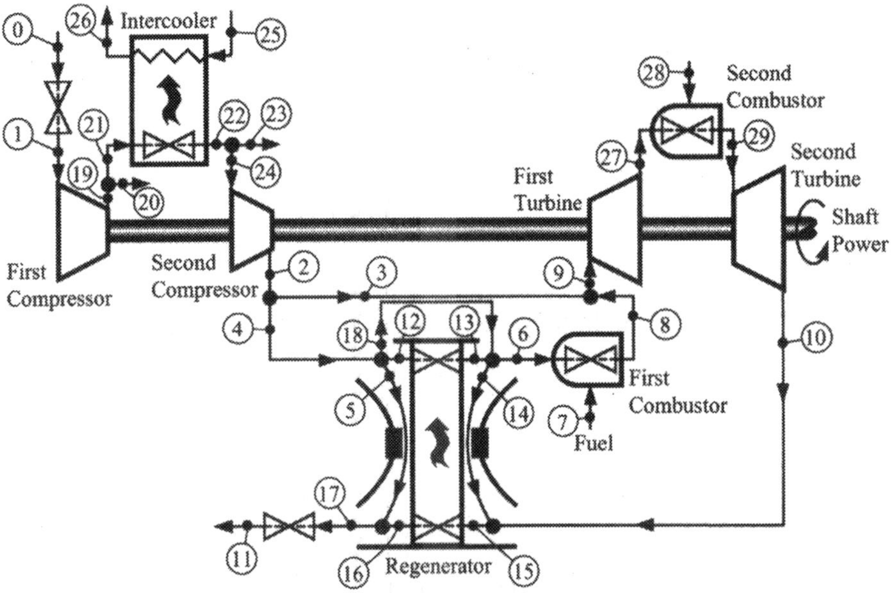

Figure 3.8: ICR Gas Turbine with Reheat.

Point 9 as it expands through the first turbine stage to Point 27. Additional fuel is burned in a second combustor to heat the gas from Point 27 to a second turbine-inlet temperature, TIT_2, at Point 29. Note that gas turbines—especially regenerative gas turbines—run fuel lean (small fuel-to-air ratios), so the gas entering the second combustor still contains enough oxygen to burn additional fuel.

Comparing the performance plots in Figures 3.7 and 3.9 shows how reheating the gas with a second combustor improves engine performance. From Equation 3.15 we know that the power produced by a turbine is proportional to the temperature drop from inlet to outlet. The sum of the temperature drops (from Point 9 to Point 27 and from Point 29 to Point 10) in the ICR gas turbine with reheat is greater than the single temperature drop (from Point 9 to Point 10) in the ICR gas turbine without reheat. Thus, the ICR gas turbine with reheat produces more power, so cycle specific power is improved.

Consider the thermal efficiency of the ICR gas turbine with reheat. The temperature of the gas leaving the second turbine stage (at Point 10) is greater than the associated temperature in the ICR gas turbine without reheat. For a high thermal efficiency, most of the available thermal energy in this exhaust must be recovered by a regenerator, or a Heat-Recovery Boiler (HRB), etc. The regenerator shown in Figure 3.8 recovers enough heat that the final exhaust

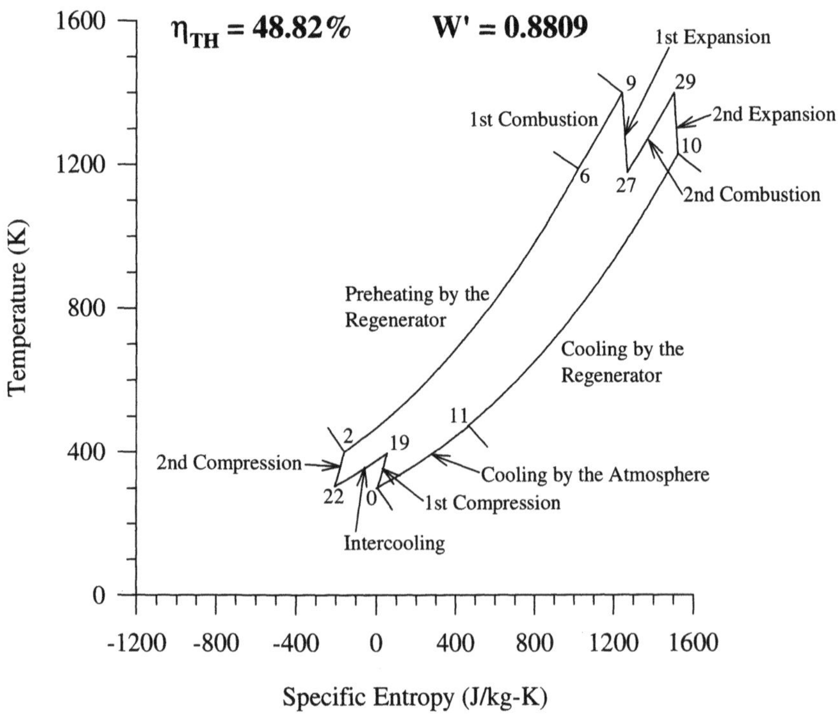

Figure 3.9: Design-Point Performance and Plot of Temperature vs. Specific Entropy for an ICR Gas Turbine with Reheat with Component Specifications Listed in Table 3.5 (data calculated by RGT-OPT).

Component Specification	Symbol	Value
Inlet Temperature of the First Turbine Stage	TIT_1	$1,400K$
Pressure Ratio of the First Compressor Stage	r_{C1}	2.25
Polytropic Total-to-Total Efficiency of the First Compressor Stage	η_{p00C1}	83%
Polytropic Total-to-Total Efficiency of the First Turbine Stage	η_{p00E1}	89%
Dimensionless Pressure Drop Across the First Combustor	$(\Delta P/P)_{H1}$	4%
Turbine-Blade-Cooling Constant	$K_{Cooling}$	0
Turbine Blade Temperature	T_{Blade}	$1,400K$
Dimensionless Inlet Pressure Drop	$(\Delta P/P)_I$	2%
Dimensionless Outlet Pressure Drop	$(\Delta P/P)_O$	2%
Dimensionless Regenerator Inlet Leakage	$\dot{m}_{XLI}/\dot{m}_4$	2%
Dimensionless Regenerator Outlet Leakage	$\dot{m}_{XLO}/\dot{m}_{13}$	2%
Atmospheric Temperature	T_{atm}	300 K
Regenerator Effectiveness	ϵ_X	95%
Dimensionless Pressure Drop Across the Compressed-Air Side of the Regenerator	$(\Delta P/P)_N$	2%
Dimensionless Pressure Drop Across the Exhaust Side of the Regenerator	$(\Delta P/P)_X$	2%
Intercooler Effectiveness	ϵ_I	95%
Dimensionless Pressure Drop Across the Air Side of the Intercooler	$(\Delta P/P)_I$	2%
Dimensionless Intercooler Inlet Leakage	$\dot{m}_{ILI}/\dot{m}_{19}$	0
Dimensionless Intercooler Outlet Leakage	$\dot{m}_{ILO}/\dot{m}_{22}$	0
Pressure Ratio of the Second Compressor Stage	r_{C2}	2.22
Polytropic Total-to-Total Efficiency of the Second Compressor Stage	η_{p00C2}	83%
Inlet Temperature of the Second Turbine Stage	TIT_2	$1,400K$
Pressure Ratio of the First Turbine Stage	r_{E1}	2.25
Polytropic Total-to-Total Efficiency of the Second Turbine Stage	η_{p00E2}	89%
Dimensionless Pressure Drop Across the Second Combustor	$(\Delta P/P)_{H2}$	4%

Table 3.5: Component Specifications for the ICR Gas Turbine with Reheat Shown in Figure 3.8.

temperature (at Point 11) is only slightly higher than the associated temperature in the ICR gas turbine without reheat. This means that most of the heat content of the fuel added to the reheat combustor is converted to additional useful power. This improves the thermal efficiency of the ICR gas turbine. Figure 3.9 shows that the added pressure drop of the reheat combustor offsets gains due to the higher temperatures, so we achieve only a slight improvement in thermal efficiency over the ICR cycle (without reheat) shown in Figure 3.7. The ICR cycle with reheat should be optimized to maximize performance. This is done in an example in Chapter 5.

In this chapter we have considered improvements to the performance of a regenerative gas turbine that can be achieved by including one intercooler and one reheat combustor. In theory, additional improvements could be achieved by including additional intercoolers and reheat combustors. Practial considerations, however, limit the attractiveness of adding multiple intercoolers and reheat combustors:

1. Adding intercoolers and reheat combustors to an engine significantly increases engine complexity and reduces the attractiveness of using a regenerative gas turbine for automotive and other applications;

2. Intercoolers and reheat combustors have pressure drops, leaks, and heat losses that detract from engine performance; and

3. The potential for improvement to engine performance decreases with each additional intercooler or reheat combustor.

Future developments may make high degrees of intercooling and/or reheat practical. Appendix A shows that the Carnot efficiency is approached by a regenerative gas turbine that includes an infinite number of intercoolers and reheat combustors (an Ericsson cycle).

3.7 Summary

This chapter has introduced gas-turbine cycles. The rest of this book focuses on regenerator design to maximize the efficiencies of gas turbines. Chapter 4 shows why regenerators are attractive heat exchangers for the gas-turbine cycles described in this chapter. Chapter 5 shows how to design the high-performance regenerators described in Chapter 4. Chapter 6 details how to calculate the performances of regenerators.

Chapter 4

Regenerator Designs

This chapter describes the state of the art in the design of regenerative gas turbines. Chapter 3 might lead one to believe that thermodynamic considerations alone dictate the designs of regenerative gas turbines. This most certainly is not the case. This chapter begins in Section 4.1 by highlighting the principal reasons (not just thermodynamic) why heat exchangers are attractive for gas turbines. Emphasis is given to the advantages of regenerators over recuperators for this duty. Then, Section 4.2 describes alternative state-of-the-art regenerator designs. Section 4.3 discusses how the regenerators can be included in high-performance gas-turbine cycles. Finally, Section 4.4 describes how regenerator design could be furthered.

4.1 Significance of Heat Exchangers in Gas Turbines

Shaft-power gas-turbine engines can be designed with or without heat exchangers. Those without are called "simple-cycle engines," and have optimum compressor pressure ratios in the range of 15:1 to 50:1, the actual optimum depending on turbine-inlet-to-compressor-inlet temperature ratios and on the accumulated losses. A large aircraft gas turbine having high-efficiency components and a high turbine-inlet temperature (TIT) will have an optimum pressure ratio of between 30:1 and 60:1, whereas a small engine with a radial-flow compressor and a necessarily uncooled turbine will have an optimum pressure ratio between 10:1 and 16:1.

The incorporation of an exhaust-gas heat exchanger can theoretically confer major advantages on a turboshaft engine. The pressure ratio for maximum efficiency falls continuously as the heat-exchanger effectiveness increases, dropping to about 2.5:1 for a high-efficiency non-intercooled cycle using typical high-efficiency components. Low-mass-flow compressors designed for low pressure ratios have higher polytropic efficiencies than those designed for high pressure ratios because both the lossy small high-pressure blading and the highly loaded low-hub-tip-ratio first stages can be eliminated. In addition, the highly regenerative low-pressure-ratio cycle is basically of higher efficiency than the simple no-heat-exchanger cycle (see Figure 4.1). A highly regenerative cycle with a turbine-inlet temperature below that of many current aircraft engines would produce a cycle thermal efficiency of close to, or perhaps above, 60%, versus about 40% for the equivalent no-heat-exchanger engine. (These well-accepted aspects are discussed by Wilson [73]). A NASA study of small-engine-component technology (SECT) confirmed that these observations apply to the rigorous weight tradeoffs of shaft-power aircraft engines. Hirschkron and Russo [33] concluded that the use of ceramic rotary regenerators was the most promising method to recover energy from the turbine exhaust to increase the overall cycle thermal efficiency of helicopter engines. However, significant improvements from current designs of regenerators would be required to reduce the leakage rate to below two percent of the compressor mass-flow rate. The study further concluded that for engines in the range of 630–1500 kW (850–2,000 shp) with 2,700 °F (1,750 K) TIT, the optimum heat-exchanger effectiveness was 90%, and the optimum compressor pressure ratio was about 8:1. (The effectiveness and associated mass of aircraft heat exchangers carry an obvious penalty. Ground-based engines have higher optimum effectivenesses of about 98%.)

Rotary ceramic heat exchangers have been used also on the Allison GT-404. In a Department of Energy (DoE) study of ceramic applications in turbine engines (CATE), the lowest leakage rate for the 400 hp GT-404 was reported to be about 4% (see Helms, et al. [29]). In a more recent study of two DoE-NASA AGT 100-kW engines built by Garrett-Ford and Allison, leakage rates for both engines were reported as being about 10% (see Helms, et al. [30]). All of these engines operated with compressor pressure ratios near 4:1. In order to meet the optimal performance of the SECT study, the regenerators for the aircraft engines studied would have to show a leakage rate of one half the lowest level reported, but at double the compressor pressure ratio. This alone explains why there has been no known effort to pursue this apparently attractive aircraft-engine concept.

4.1.1 Leakage

Figure 4.2 shows the effect of leakage on the calculated thermal efficiency of a 10-kW automotive gas turbine for a hybrid engine, having a relatively low TIT (1,250 K) and low component efficiencies but a high-effectiveness regenerator. The penalty for leakage is high in such an engine; it is even higher in an engine with a higher TIT and components of higher efficiency. Therefore it is highly desirable to reduce leakage as far as possible.

In rotary regenerators, leakage occurs in two ways. The first is direct leakage through the seals that separate the high-pressure section of the heat exchanger from the low-pressure section. The second, called carry-over leakage, occurs because high-pressure fluid trapped in the core passages is carried past the seals to the low-pressure side. This problem worsens with increased core speed, which in turn is required if the designer chooses to specify a highly porous core with

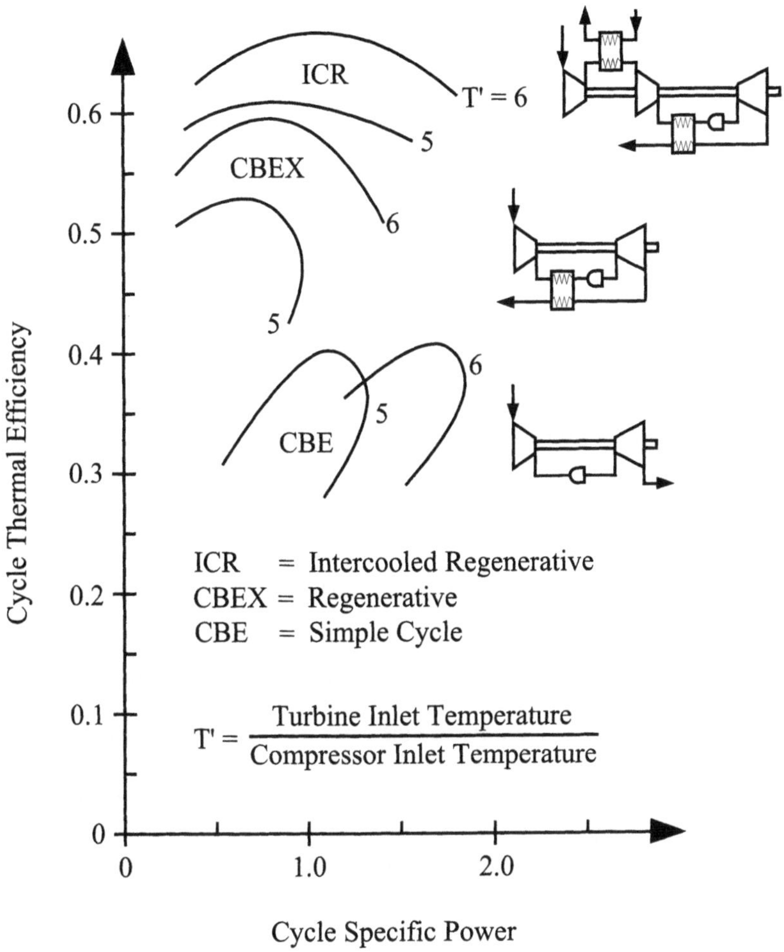

Figure 4.1: Design-Point Performance of Gas-Turbine Cycles. This material has been reproduced from the Proceedings Part A Issue A3 1993 Volume 207 p. 196 Figure 3 by David Gordon Wilson by permission of the Council of the Institution of Mechanical Engineers, London.

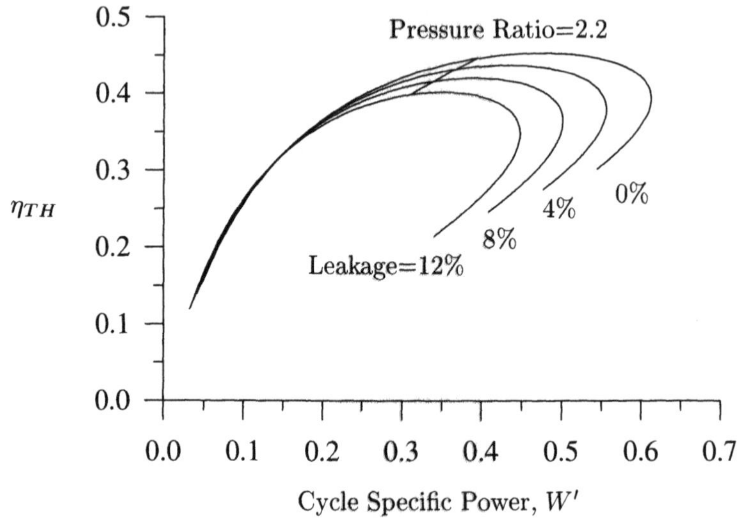

Figure 4.2: Effect of Leakage on Thermal Efficiency of a Regenerative Cycle (adapted from Wilson and Beck [81]).

a low thermal mass to reduce the overall size and weight of the heat exchanger. However, carry-over leakage is usually in the range of 0.5–1.0% of the compressor mass flow, and is therefore less of a problem than is the direct seal leakage.

4.1.2 Regenerator Advantages over Recuperators

Regenerators in general have advantages over recuperators for gas-turbine engines. One of these is compactness. In laminar flow, the volume needed for a given quantity of heat to be transferred in a convective heat exchanger is proportional to the square of the hydraulic diameter of the passages used (see Wilson [73] and Section 5.1). The passages in gas-turbine rotary regenerators can be made much smaller than those of tubular or plate-fin heat exchangers because hot- and cold-stream separation is controlled by the seals rather than by the actual exchanger material (as in recuperators), and the cost of making small passages is low.

Another desirable feature for the regenerator is low pressure drop. The pumping power required to force gas through a heat exchanger is directly proportional to the square of the Mach number and is rather independent of matrix geometry (see Wilson [73]). Therefore, large face areas must be used to minimize fluid velocity. In the rotary regenerator, elaborate manifold schemes to interleave the fluids are unnecessary, so a large flow area is practical. With fixed-surface heat exchangers, achieving both compactness and large, interleaved flow areas simultaneously is more difficult.

The principal advantage of the ceramic-core regenerator for non-intercooled-cycle engines is, however, its temperature capability. A non-intercooled regenerative cycle using modern TIT's and a reasonably high-effectiveness regenerator will have an optimum pressure ratio low enough for the turbine-outlet (regenerator inlet) temperature to be well above the maximum tolerable for a metallic heat exchanger. The principal reason for the new interest in the intercooled regenerative cycle is that the use of an intercooler increases the optimum overall pressure ratio from a range of 2.5–4:1 to a range of 5–16:1. At these higher levels the increased turbine expansion ratio brings the regenerator-inlet temperature down to the point where metallic recuperators and regenerators can be used.

4.1.3 Size Limitations for Ceramic Rotary Regenerators

Whereas metallic Ljungstrom rotary air preheaters are made with extremely large diameters (such as 10 m), the diameter of a monolithic ceramic regenerator core is limited to about 1 m because of manufacturing and thermal-stress restrictions. The original Corning cores were wound from two extruded ribbons: one smooth and one wavy. The resulting core has near-triangular passages, each having one rounded corner and two cusped corners. The trend is to make cores by extruding the whole face through a die. This allows the use of alternative channel cross-sections, such as the rectangular geometry.

The manufacture of acceptable whole-face extrusions becomes more difficult the larger is the diameter of the disk. However, before a manufacturing diameter limit is reached, a limiting diameter is attained at which crack-growth rate from thermal stresses predominates. As Figure 1.4 shows, one face of the core heats up to a temperature close to that of the gases leaving the turbine (which is likely to be above 1,050 K), while the other face is close to the compressor-delivery temperature (which could be below 400 K). Thus thermal stresses, even in materials with very low thermal expansion coefficients, can be high, especially if a low pressure drop is specified, because this will require a thin (short-axial-length) core.

Corning, and presumably others, have produced ceramic cores with diameters of over 2 m by cementing together separate core pieces. This would not, however, overcome the thermal-stress problem. Perhaps a more promising approach to the development of a large ceramic regenerator would be to produce the disk in separate ceramic sector modules. This line of thinking produced the modular-regenerator concept described below.

4.2 Alternative Regenerator Designs

4.2.1 Two-Chamber and Rotary Regenerators

There is a subtle but important difference between switchable regenerators of an open-hearth furnace or a Cowper hot-blast stove and a rotary regenerator. A switchable regenerator is first heated up until the core material is almost uniformly hot. The flow is then switched so that cold air passes through the

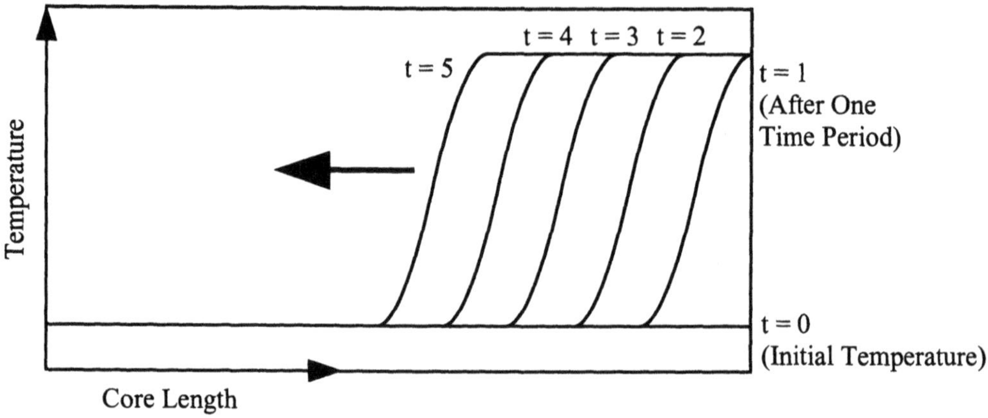

Figure 4.3: Propagation of a Thermocline through the Core of a Switching Regenerator. This material has been reproduced from the Proceedings Part A Issue A3 1993 Volume 207 p. 197 Figure 4 by David Gordon Wilson by permission of the Council of the Institution of Mechanical Engineers, London.

core, and that continues until the core is almost uniformly cold. Figure 4.3 shows how a thermocline propagates through the core during this process.

In contrast, in the rotary regenerator, the ideal case is for the core temperature to change very little as it rotates. For this to happen, suppose that the core has first reached a steady-state linear temperature distribution, as Figure 1.4 shows. If the core is rotated very fast so that only an insignificant mass of hot gas has time to flow when it is open to that flow, and likewise only a vanishingly small mass of cold flow passes through the channels, the core material would not change temperature significantly. In practice, the rotation rate may be set according to the recommendations of Kays and London [37] to give a solid thermal-capacity rate (C_R—the product of solid mass flow and mean specific heat) of five times that of the fluid with the lower thermal capacity rate (C_N). This is a close-enough approach to an infinite rotation rate. (Hagler and Wilson [23] recommend that an even lower value, $C_{ROT} \equiv C_R/C_N = 3$, may be nearer to an optimum for many gas-turbine regenerators. The optimal value is $C_{ROT} = 4.3$ for the ICR gas turbine with reheat considered in Section 5.2.) The core-temperature distribution will then change by a small amount during rotation through each fluid, but it will still resemble that of a counterflow recuperator, and a high effectiveness can still be attained. The leakage, however, is a rather separate issue.

4.2.2 Alternative Rotary-Regenerator Configurations

The circular disk, with or without a central hole, with axial flow of the two fluids, is the most-common form of rotary regenerator (see Figure 4.4). The cylindrical-core configuration, with radial flows has also been proposed and used. Figures 4.4 and 5.1 show overall configurations of radial-flow regenerators.

An intermediate form is to configure the core as a polygon of planar honeycomb sections. The regenerator core can also be formed of, for instance, sintered balls, perhaps with some radial partitions to prevent circumferential flow when the pressure difference across the longitudinal seals would otherwise induce leakage.

There could be some advantages in particular circumstances for intermediate configurations of regenerator, for instance conical, but the manufacturing costs would seem to be high for such shapes.

4.2.3 New Regenerator Concepts

If the preceding discussion is accepted, two development needs stand out for regenerators. In automotive sizes, the seal leakage and wear rates must be greatly reduced. In the large sizes for which thermal stresses and other manufacturing considerations make rotary ceramic regenerators feasible, mechanisms must be developed that permit ceramic cores to operate in very large sizes, perhaps initially for engines of 10–100 MW, and that enable maintenance and replacement activities to be carried out without engine shut-down.

4.2.4 Discontinuous-Rotation Regenerator

The principal problem in sealing rotary regenerators is that of maintaining close-contact seals. In general, rubbing seals are used. The wear on seals and on the delicate cell walls can be high, and leakage rates normally increase, sometimes rapidly, during operation. A simple early concept of operating a rotary regenerator was to hold the matrix stationary while exchanging heat with the two fluids, and then to rotate it rapidly through 180 degrees. In this form it would operate as a two-chamber Cowper stove. However, the valves would be less effective than those fitted on stoves. The valve or seal leakage would be reduced if the seals were arranged to clamp to the matrix when it was stationary, and to lift slightly when the matrix were rotated (see Wilson [75]). In such an arrangement the matrix can be moved incrementally by much less than 180 degrees. Beck has shown that a series of angles of incremental rotation will produce no penalty to heat-transfer performance over continuous rotation whatever is the angle chosen between the radial seals (see Beck [6] and Section 6.1.13). One can choose to have the seals bounding the compressed-air passage at, for instances, 135° or 150°, in which case there is no heat-transfer penalty in rotating the matrix through increments of 45° or 30°. Figure 4.5 illustrates diagramatically the discontinuous-rotation regenerator. Its operation is as follows. The core is stationary with the seals clamped (through the action of linear actuators) for a period of perhaps two seconds. The seals are then withdrawn perhaps 0.05 mm (0.002"), the matrix is rapidly rotated by a mechanism through 30°, and the seals are clamped again. This movement should be accomplished in 0.1

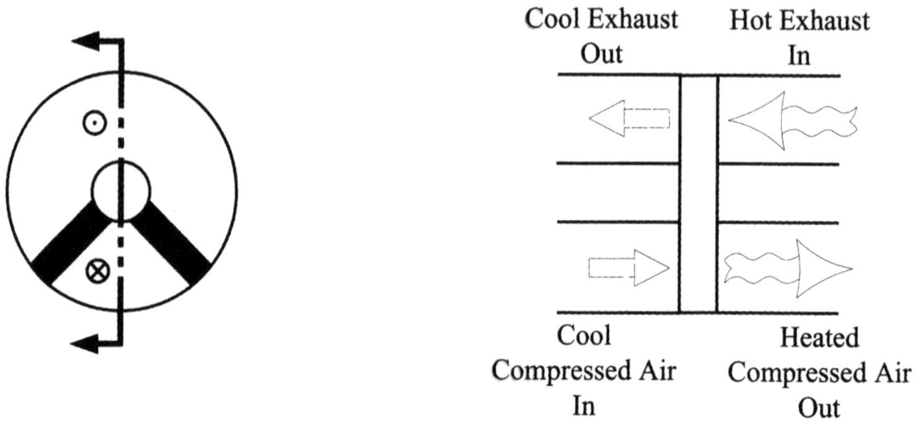

(a) Circular Disk with Axial Flows

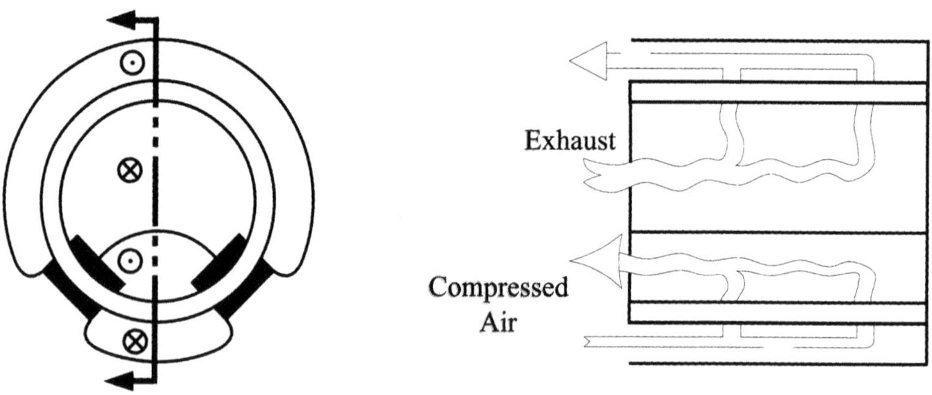

(b) Cylindrical Core with Radial Flows

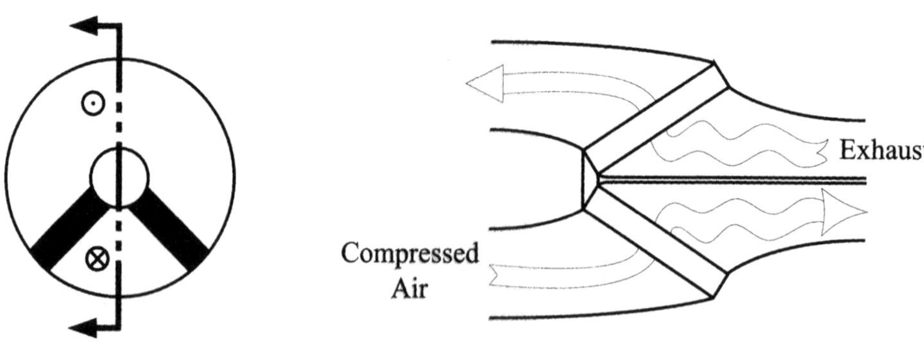

(c) Conical Core with Axial-Radial Flows

Figure 4.4: Alternative Forms of Rotary Regenerators.

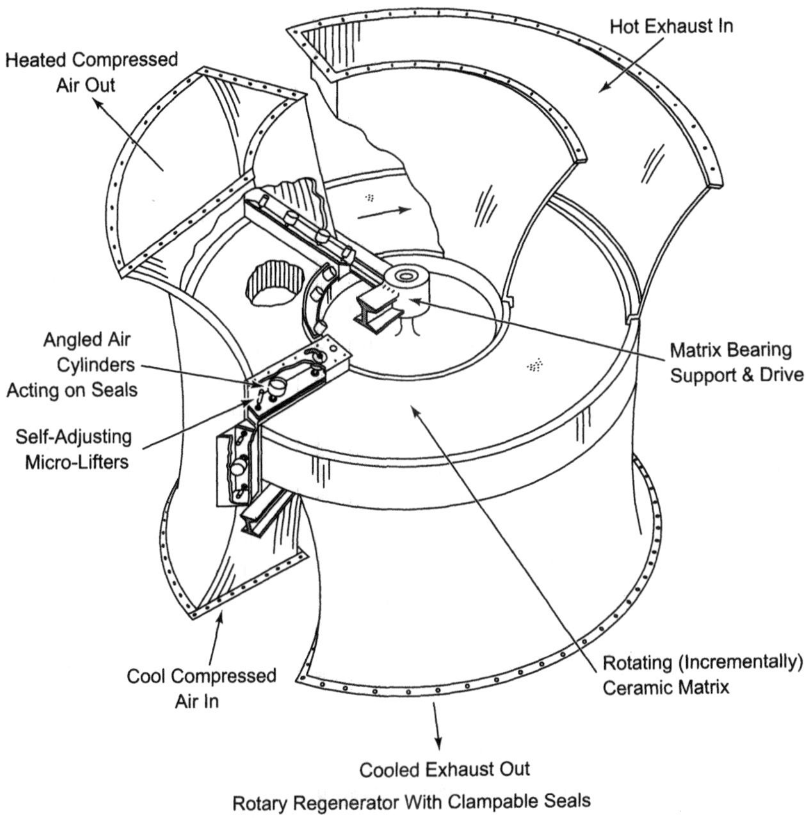

Figure 4.5: Discontinuously Moving Regenerator with Clampable Seals. This material has been reproduced from the Proceedings Part A Issue A3 1993 Volume 207 p. 200 Figure 8 by David Gordon Wilson by permission of the Council of the Institution of Mechanical Engineers, London.

seconds. Seal leakage could consequently be reduced to 5% of the continuous-rotation value. A reduction of anything approaching this magnitude would have a major favorable effect on the thermal efficiency and specific power of engines designed with rotary regenerators without adding more than a small degree of complexity in the matrix-drive and seal-adjustment mechanisms. Moreover, wear on the seals and matrix would be reduced to a negligible level.

Such a solution would be limited to smaller engines. Another approach is required if regenerators are to be used in larger gas turbines.

4.2.5 Modular Regenerator

Present rotary regenerators use face seals on ceramic cores that have solid ceramic centers and rims. The seal lengths are long, and the seals act against very slow core movement (of the order of 10–100 mm/s). Leakage would be reduced if the total seal length were reduced. This would mean that instead of a long seal acting over a slow-moving core, a short seal would be associated with a faster-moving core. The obvious direction in which to go is to employ an annular core (see Figure 4.6). The leakage across the circumferential seals acting on smooth solid rims is small compared with the leakage across the radial seals acting on the open structure of the core. Increasing the inner diameter decreases the length of the radial seals. At some point in this sequence it would be more appropriate to re-locate the seals so that they surround the core as in Figure 4.6, because they would be considerably shortened. There would be complications on the rotor-drive arrangements.

A logical development is to break such an annular core into small parts (modules) to reduce both thermal stresses and manufacturing costs. In this case, there would be little point in retaining the modules as arcs. Making the modules in the form of rectangular solids endows the regenerator with many desirable attributes at one stroke. Modules of this shape can be assembled to form rectangular flow passages if the modules are moved transversely into the heat-transferring chamber (see Figure 4.7). The seals could be clampable as in the previous concept. However, for a modular regenerator, there is little penalty in using large sealing lengths, and doing so avoids the complication of seal-clamping mechanisms. We are, therefore, treating the closely spaced transfer-duct walls as long labyrinths. Seal leakage is predicted, therefore, to be low.

The complete regenerator consists of two heat-transferring chambers, one in the compressor-delivery stream where low seal leakage is important, and one in the exhaust, where leakage has less consequence. The face area of the exhaust chamber would be two-to-three times that of the compressed-air chamber if it was desired to have similar percentage pressure drops. The modules could be moved over a distance, so that some expensive lengths and bends in the compressor-delivery or exhaust ducts could be eliminated. In the length of transfer duct between the two heat-exchanging chambers it would be practicable to divert modules out of service for periodic maintenance or replacement. The cost in high-volume production of a full complement of standard modules would be far less than that for manufacturing a large monolithic rotor. It is, therefore, a question of straightforward engineering economics whether, for some special applications such as exhaust-heated coal-burning gas turbines, modules in which high-temperature alkali deposition had occurred should be simply sacrificed, or whether the development of ceramic materials that could resist alkali attack at high temperatures would be justifiable and desirable.

4.3 Gas-Turbine Cycles with Heat Exchangers

This section puts the Brayton cycles described in Chapter 3 in the context of regenerator design. Alternative gas-turbine cycles that include heat exchangers are described with the focus of how these cycles can be effected with the state-of-the-art regenerators described in the previous sections.

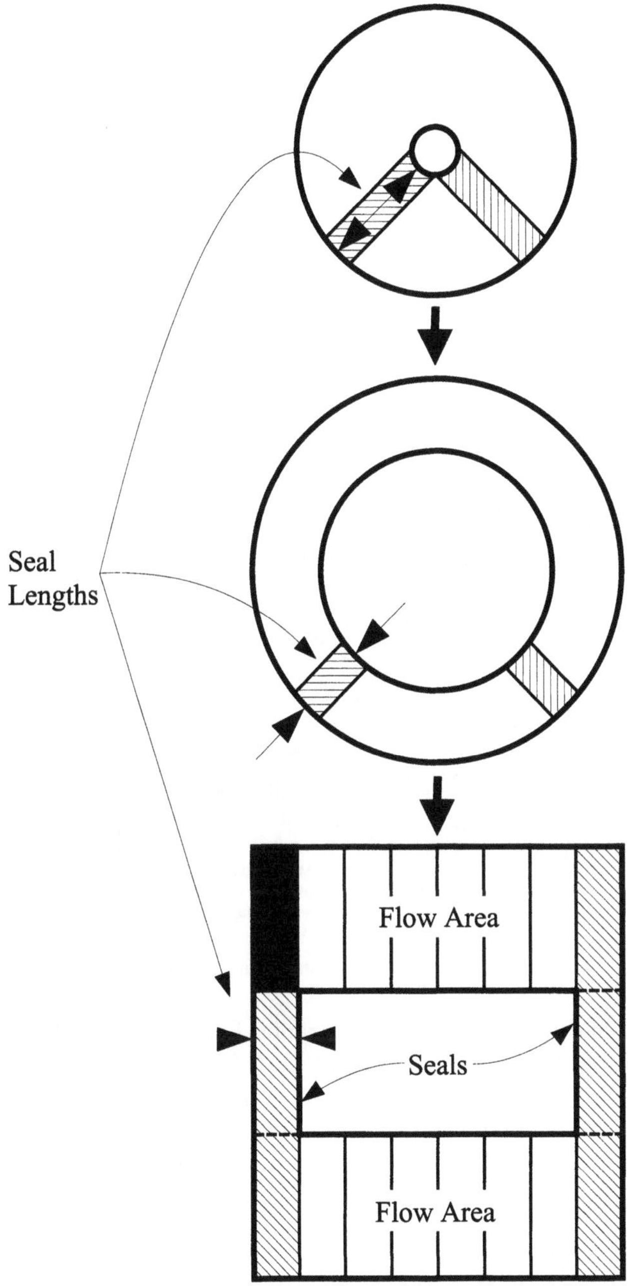

Figure 4.6: Sequence from Rotary-Disk to Linear Modular, Showing Reduction in Seal Length.

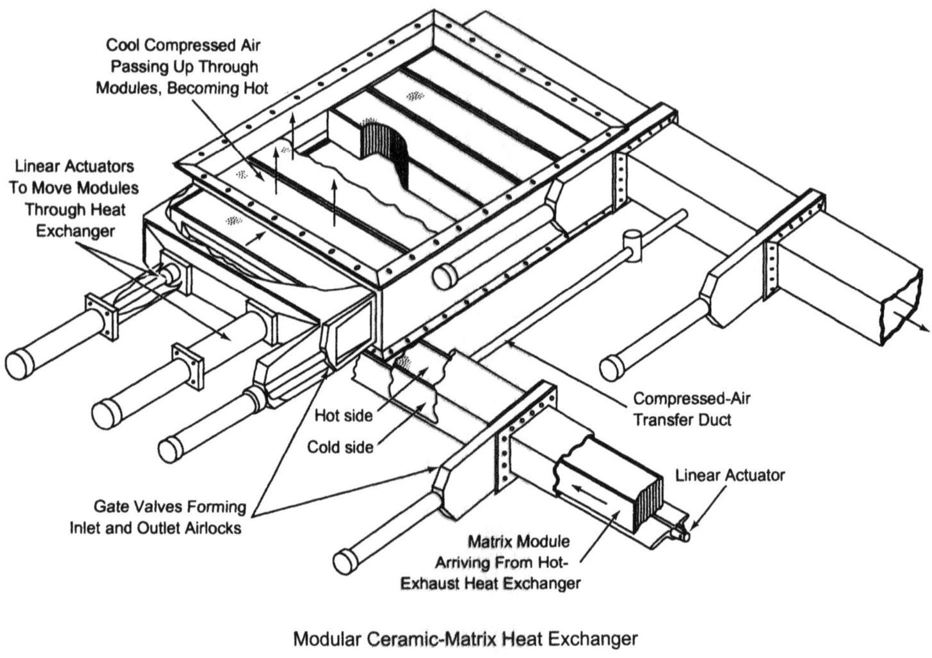

Figure 4.7: One Form of Modular Ceramic Regenerator. This material has been reproduced from the Proceedings Part A Issue A3 1993 Volume 207 p. 200 Figure 10 by David Gordon Wilson by permission of the Council of the Institution of Mechanical Engineers, London.

4.3.1 The "Simple" Regenerative Cycle

This cycle has been used for almost all automotive engines. Its cycle designation is CBEX, for "Compressor, Burner, Expander, heat eXchanger." When recuperators are used with this cycle as, for instance, in the Lycoming AGT 1500 in the US-Army M1 tank, rather low effectivenesses have been chosen, e.g. 68% in this engine, because of the extremely limited volume allowed for the recuperator (see Wilson and Kupratis [85]). The resulting optimum pressure ratio is about 14:1. A regenerator fitted into the same space limits might have an effectiveness of 90 or 95%, in which case the optimum pressure ratio is likely to be 3 or 4:1, the compressor and turbine efficiencies will be higher at design point and over a broader operating range, and the full-load and part-load efficiencies will be much higher. A vital question for automotive engines, which are always volume limited, is whether or not the lower specific power of low-pressure-ratio cycles will be counterbalanced by the higher specific power resulting from the higher compressor and turbine efficiencies. In some unpublished cases examined by the authors the specific power was almost unchanged because these two effects cancelled each other.

If the engine has metal turbine blades large enough to be cooled, which would be the case for engines above about 1 MW, the increased turbine-inlet temperature that cooling makes possible leads to higher turbine-outlet (heat-exchanger-inlet) temperatures. These temperatures are too high for metallic recuperators. This size of engine, 1 MW, is also near the upper limit at which twin-disk regenerators can be used (because of the size limit for manufacturing the disks). This is the reason for the restriction (up to the mid-1990's) of simple regenerative cycles to small-size low-pressure-ratio units with high-effectiveness (90% and over) rotary ceramic regenerators and metallic or ceramic blading, higher pressure ratios, and low-effectiveness (e.g. 80% and below) metal recuperators. In the mid-1990's there are underway several attempts to produce ceramic recuperators that would push higher the recuperator temperature limit, but in the authors' opinion they will be useful only if they can be produced in high-effectiveness versions, so that they can convey the advantages of high design-point and part-load efficiencies and low pressure ratios that come with the optimization of high-effectiveness regenerators. We believe (with possible but understandable bias) that the modular ceramic regenerator could provide a more-promising route to the exploitation of the simple regenerative cycle to higher power levels.

4.3.2 The Intercooled-Regenerative Cycle

The intercooled recuperative (ICR) cycle, having the component designation CICBEX, has several advantages over the non-intercooled heat-exchanger cycle. Gas turbines produce net power as the difference between turbine output power and compressor input power. (The compressor input power might be one-half to three-quarters of the turbine output power [see Chapter 3].) The compressor power input is considerably reduced by intercooling. Therefore, the percentage increase in net power is large. If the heat-exchanger effectiveness is high, intercooling brings about very little increase in fuel requirements, so that the efficiency also increases.

The optimum pressure ratio for each of the two compressors is found, on analysis, to be close to the value for the single compressor of a non-intercooled heat-exchanger cycle (see Figure 4.8). Therefore, the overall pressure ratio is reasonably high, as is the expansion ratio for the turbine. The temperature drop through the turbine is thereby increased, and a metallic recuperator can be used. This property of the ICR cycle allows it to be used in the 20-MW Westinghouse/Rolls-Royce WR-21 for marine propulsion (see Mullins [54]; Hawkins, et al. [28]; Shepard, et al. [66]; and Chiprich [13]) and in the Westinghouse Advanced Turbine System (see Bannister, et al. [3]). The non-intercooled heat-exchanger cycle cannot at present be used for high-power engines, because ceramic regenerators are not produced for engines above about 1 MW.

4.3.3 The Exhaust-Heated Cycle

The exhaust-heated cycle was developed by Donald Mordell in the 1950's as a means of burning coal in a gas turbine without the products of combustion passing through the turbine blading (see Mordell [52]). A re-examination of the cycle using presently available materials and including the consideration of using a ceramic-core regenerator instead of the tubular metallic heater employed by Mordell was carried out by Wilson, et al. [79]. The results were generally favorable. We show an advantageous variation of this cycle (the supplementary-fired exhaust-heated cycle) in Figure 4.8, with some design-point efficiencies (see Wilson [82]).

For a maximum thermal efficiency, a low pressure ratio and a high turbine-inlet temperature would be used, which would require a ceramic heat exchanger with high-temperature capability such as the Modular Regenerator described above. However, low-pressure-ratio gas turbines adaptable to include a heat exchanger are not currently produced. Therefore, turbine-outlet temperatures are low, and a significant amount of heat can be added in the secondary combustor while maintaining the heat-exchanger-inlet temperatures below the limit for metallic heat exchanagers.

4.3.4 Solar-Heated Cycle

A three-chamber modular ceramic regenerator as shown diagramatically in Figure 4.9 would be appropriate for a solar-heated gas turbine working on the simple regenerative cycle. The heat-exchanging face through which the turbine-exhaust flow passes (at near atmospheric pressure) would be as before (see Figure 4.7). The modules would then move to a half-height face, also at atmospheric pressure. This would be open to the directed rays from the solar mirrors or other forms of concentrator. The deep passages would be ideal traps for the solar radiation. The modules would then be moved into the compressor-delivery face. The design-point performance would be that of an exhaust-heated cycle. It

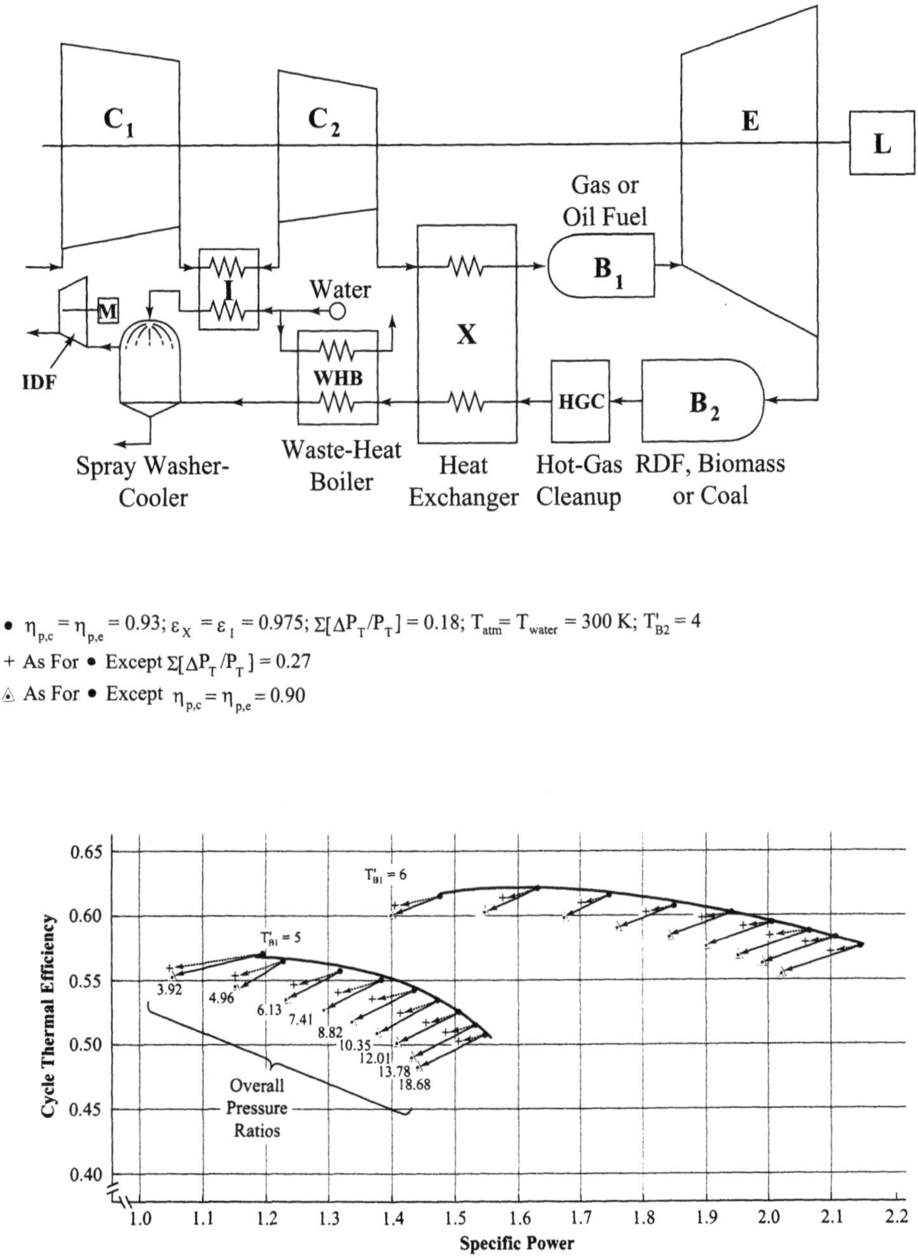

Figure 4.8: The Modular Regenerator in the Supplementary-Fired Exhaust-Heated Cycle with Design-Point Performance. This material has been reproduced from the Proceedings Part A Issue A3 1993 Volume 207 pp. 205 and 206 Figures 5 and 6 by David Gordon Wilson by permission of the Council of the Institution of Mechanical Engineers, London.

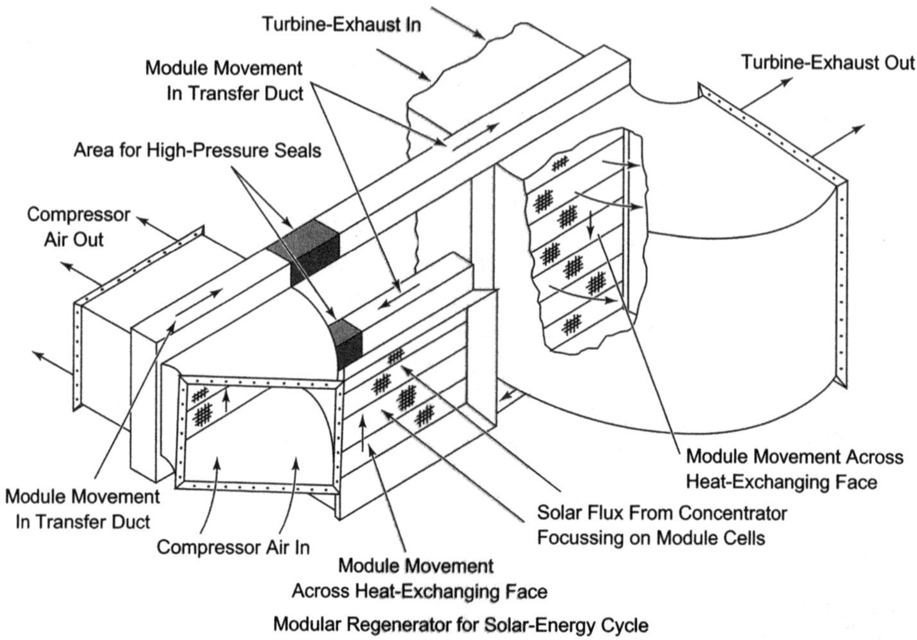

Figure 4.9: Modular Regenerator Used in a Solar-Heated System.

should be expected that the temperature to which the modules could be heated would be higher than that for the coal-or-biomass-heated cycle because of the absence of concern for the effects of alkalis and other deleterious products of combustion.

4.4 Future Directions

Regeneration has received very little public funding relative to the massive funding that has been devoted to the improvement of all other processes in gas-turbine engines over the last half century. The pioneering efforts of people who have worked in the past with mostly private corporate funds must be applauded. It is to be hoped, however, that so large a field that is so potentially rewarding environmentally will receive appropriate public funding in several countries. There are many avenues awaiting exploration, of which those touched on in this book are merely a selection.

Chapter 5

Design Procedures and Examples

This chapter describes how to design high-performance gas-turbine regenerators. The goal of designing a regenerator is to calculate the required dimensions to satisfy specifications. We describe three alternative methods for calculating these dimensions, and demonstrate the methods with example regenerator designs. First Section 5.1 describes Direct Regenerator Design, a method for designing regenerators directly from the designer's specifications. With this method, the designer specifies regenerator parameters (for example effectiveness) as well as parameters associated with the gas-turbine cycle (for example the inlet temperatures to the regenerator). From these specifications the designer directly calculates the size and weight of the regenerator's core as well as its rotation period. The advantage of this first method is its simplicity and directness. However, Section 5.2 shows that an alternative design method, Optimal Regenerator Design, leads to optimal regenerators that maximize cycle thermal efficiency. These first two design methods have been included in the commercial software RGT-OPT™ [1]. Sections 5.1 and 5.2 include four regenerator design examples and output from RGT-OPT. We have included in Section 5.3 a design method that is attributed to Kays and London [37] and is advocated by the second author. We illustrate this method with two examples.

5.1 Direct Regenerator Design

This section describes Direct Regenerator Design, a method for designing regenerators directly from the designer's specifications. We illustrate this design method with three examples:

[1] RGT-OPT is a trademark of Douglas Stephen Beck, the principal author, and is available from him.

1. A Mesh Screen Matrix (MSM)[2] regenerator for a 20-MW power-generation gas turbine ;

2. A regenerator with parallel-plate passages for a 3-MW rotorcraft gas turbine; and

3. A regenerator with a ceramic-disk core for a 75-kW automotive gas turbine.

First we discuss specifications for Direct Regenerator Design in the next section. Then, in Section 5.1.2 we describe a five-step procedure for calculating regenerator dimensions from the specifications.

5.1.1 Specifications and Results

Table 5.1 lists parameters the designer specifies in Direct Regenerator Design. The specifications are for the three applications listed above. Figures 5.1 and 5.2 show schematics that define the dimensions of the three regenerator types . Table 5.2 lists the dimensions of the regenerators that result from this design procedure.

Cycle Data. The cycle data listed in the top portion of Table 5.1 result from specifications for the regenerative gas turbine for which the regenerator is being designed. In Direct Regenerator Design, we treat these data as fixed. However, we could iterate by alternately calculating the cycle and designing the regenerator. This is what we do in Optimal Design to find the gas-turbine design *and* regenerator design that maximize thermal efficiency.

Regenerator Effectiveness. We have specified effectivenesses of 95% for the regenerators of the 20-MW power-generation gas turbine and the 75-kW automotive gas turbine. This effectiveness value is currently achieved by regenerators and is typically close to optimum for land-based power generation and for automobiles: lower effectivenesses reduce cycle performance because not as much heat is recovered from the exhaust, and higher values lead to large regenerators with high leakages, pressure drops, and power consumption. We have chosen an effectiveness of 90% for the 3-MW rotorcraft application because Hirschkron and Russo [33] found this value leads to a good balance for rotorcraft of engine efficiency, size, and weight.

Core Volume Ratio, V' . The core volume ratio, V', is defined as the ratio of the core volume on the compressed-air side to the core volume on the exhaust side. We specify a value of $V' = 0.3$ for the regenerators of the 20-MW power-generation gas turbine and the 75-kW automotive gas turbine and $V' = 0.4$ for the rotorcraft application. A core volume ratio of $V' = 0.3$ means that the regenerator volume on the exhaust side is about three times larger than

[2] See Kays and London [37] for details of Mesh Screen Matrix (MSM) regenerator cores.

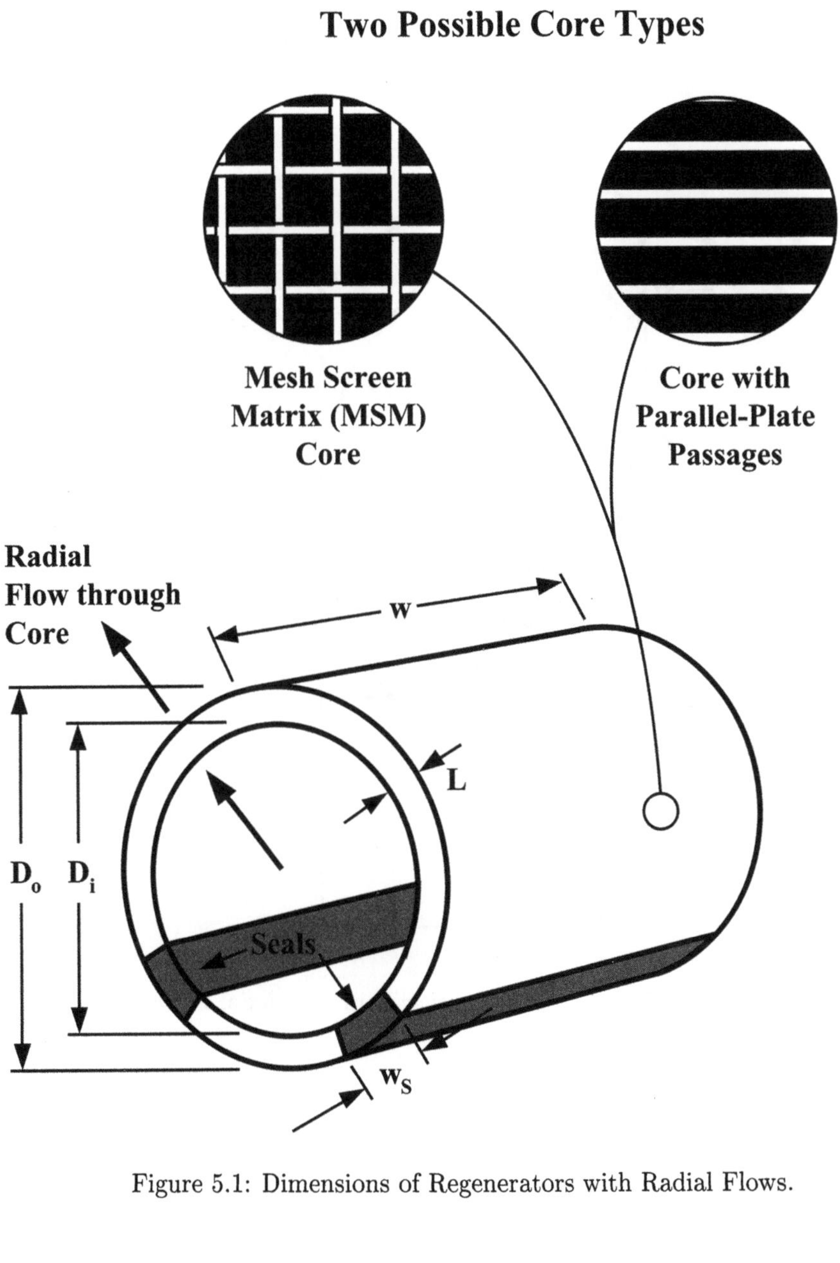

Figure 5.1: Dimensions of Regenerators with Radial Flows.

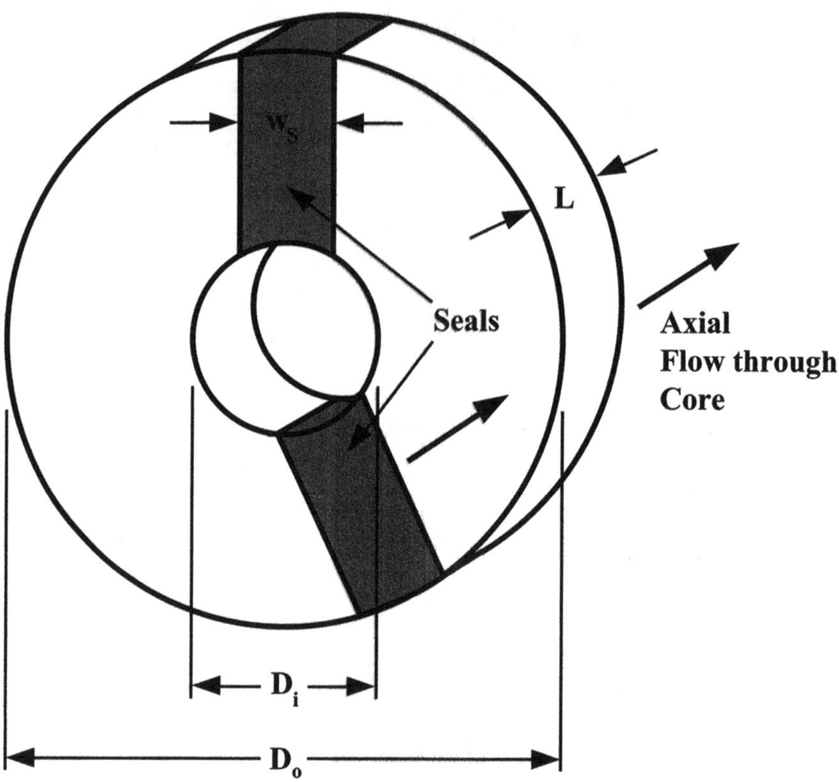

Figure 5.2: Dimensions of Regenerators with Axial Flows.

Spec.	Units	Description	20-MW GT	3-MW GT	75-kW GT
Cycle Data:					
T_{Ni}	K	Inlet Temperature on the Compressed-Air Side	458.34	641.34	439.05
T_{Xi}	K	Inlet Temperature on the Exhaust Side	890.82	1,074.82	1,139.12
f_X	–	Fuel-to-Air Ratio on the Exhaust Side	1.801%	1.949%	0.8559%
$\dot{m}_N$	kg/s	Air Mass Flow Rate	47.87	7.641	0.4518
$\dot{m}_X$	kg/s	Exhaust Mass Flow Rate	49.31	8.027	0.4557
P_{Ni}	MPa	Inlet Pressure on the Compressed-Air Side	1.1770	0.98020	0.29406
P_{Xi}	MPa	Inlet Pressure on the Exhaust Side	0.10408	0.10408	0.10408
$(\Delta P/P)_X$	–	Dimensionless Pressure Drop Across the Exhaust Side	0.02	0.02	0.02
Regen. Specs.:					
ϵ	–	Regenerator Effectiveness	0.95	0.90	0.95
V'	–	Core Volume Ratio, the Ratio of the Volume on the Compressed-Air Side to the Volume on the Exhaust Side	0.3	0.4	0.3
C_{ROT}	–	Dimensionless Core Rotation Rate	3.0	3.0	3.0
SC	–	Seal Coverage, the Fraction of the Core Face Covered by Seals	0.1	0.1	0.1
δ	mm	Seal Clearance	0.5	0.5	0.2
K_{LEAK}	–	Seal-Leakage Coefficient	0.6	0.6	0.6
D_i	m	Inner Diameter of the Core	2.0	1.0	0.05
p	–	Core Porosity, the Ratio of the Voids Volume to the Total Volume	0.8	0.85	0.8
Nu	–	Nusselt Number	3.61	8.235	4.12
C_f	–	Friction Coefficient	14.2	24.0	15.6
D_H	mm	Hydraulic Diameter	0.3	0.5	0.5
ρ_R	kg/m^3	Density of the Core Material	7,770	7,770	2,260
c_R	J/kg-K	Specific Heat Capacity of the Core Material	460	460	1,100
k_R	W/m-K	Thermal Conductivity of the Core Material	25	25	2.7

Table 5.1: Specifications for Direct Regenerator Design for Three Applications.

Dimension	Units	Description	20-MW GT	3-MW GT	75-kW GT
w	m	Core Width	3.984	1.019	–
τ_{ROT}	s	Rotation Period	3.417	1.712	6.651
L	mm	Core Thickness or Flow Length	28.83	24.68	68.95
m_R	kg	Mass of the Core	1,138	94.35	8.937
w_S	m	Seal Width	0.3187	0.1610	0.05155
D_o	m	Outer Diameter of the Core	2.058	1.049	0.6063
$(\Delta P/P)_{Nc}$	–	Dimensionless Pressure Drop Across the Core on the Compressed-Air Side	0.04572%	0.04703%	0.7459%
$\dot{m}_{XLI}/\dot{m}_4$	–	Dimensionless Regenerator Inlet Leakage	1.409%	1.943%	1.281%
$\dot{m}_{XLO}/\dot{m}_{13}$	–	Dimensionless Regenerator Outlet Leakage	1.438%	1.983%	1.308%

Table 5.2: Regenerator Dimensions that Result from the Data Listed in Table 5.1.

the volume on the compressed-air side, so the frontal area on the exhaust side is also about three times larger. The exhaust has a lower pressure than the compressed air, it is hotter, and it has a slightly higher flow rate. These factors mean that unless we make the flow area on the exhaust side much larger than the flow area on the compressed-air side, the pressure drop across the core on the exhaust side will be much larger than the pressure drop across the core on the compressed-air side. In fact, the pressure drop across the core on the compressed-air side will typically be too low for good flow distribution across the core face, even if the designer makes the flow area on the exhaust side three or four times larger than the flow area on the compressed-air side. For good flow distribution on the compressed-air side, the designer should include a partial screen to increase the pressure drop to about 1–2%, as advocated by the second author. The designer should not make the flow areas too dissimilar because regenerators have minimum core volumes if the convective conductances on each side are equal (if the convective-conductance ratio $(hA)' = 1$), and the size of the core increases with the degree to which the convective conductances are dissimilar. In practice, the designer should select a core volume ratio in the range $0.2 \leq V' \leq 0.6$ so that the high-pressure containment for the compressed air is small, but the overall size of the regenerator's core is not too large.

Dimensionless Core Rotation Rate, C_{ROT} . We have chosen dimensionless core rotation rates of $C_{ROT} = 3$ for all three applications. This choice results from a trade-off of high effectiveness (fast core rotation) and low carry-over leakage (slow core rotation). Figure 5.3 shows the effect that core rotation rate has

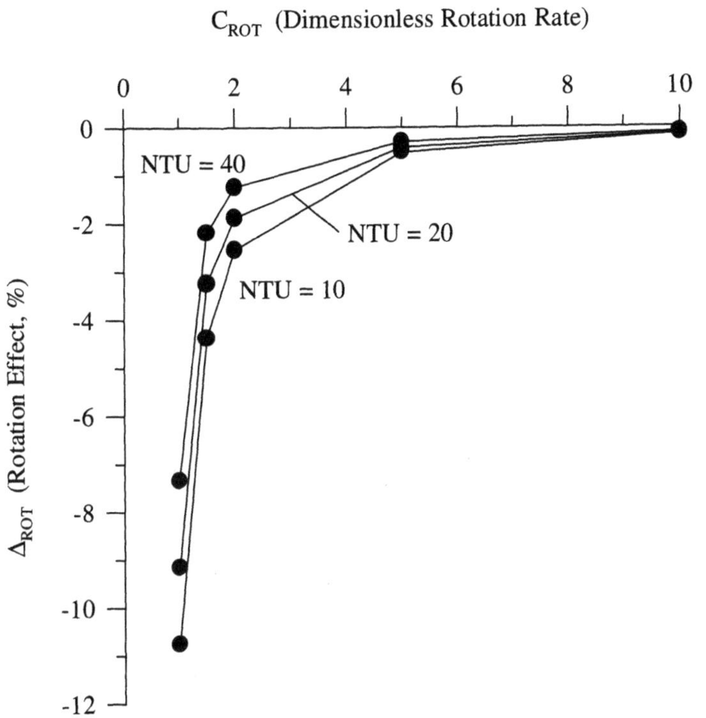

Figure 5.3: Effect of Core Rotation Rate on Regenerator Effectiveness (data from Bahnke and Howard [2]).

on regenerator effectiveness. The y-coordinate in Figure 5.3 is Core-Rotation Effect :

$$\Delta_{ROT} \equiv \frac{\epsilon_{ROT} - \epsilon_{CF}}{\epsilon_{CF}}, \tag{5.1}$$

where

Δ_{ROT}	=	Core-Rotation Effect (–);
ϵ_{ROT}	=	Effectiveness of a regenerator with a finite core rotation rate (–); and
ϵ_{CF}	=	Effectiveness of a counter-flow heat exchanger.

For very fast rotation rates the temperature profile through the core changes very little during each rotation, and the effectiveness approaches the effectiveness of a counterflow heat exchanger with identical heat-transfer parameters. For slower rotation rates, the temperature profile through the core fluctuates significantly during each rotation and effectiveness is reduced. Figure 5.3 shows an elbow at a dimenionless core rotation rate of $C_{ROT} \approx 3$. The Core-Rotation Effect at this rotation rate is small. Below this value, effectiveness is significantly reduced. Benefits of faster rotation rates are small.

Seal Coverage, SC. For all three applications we have chosen a Seal Coverage of $SC = 10\%$. To minimize leakage, we should make the seals that separate the flows wide, so we should specify a large Seal Coverage. However, for a given total core volume, large Seal Coverage reduces the percentage of the core volume that is active (exposed to the flows at any given instant of time). In addition, large Seal Coverage reduces effectiveness for the regenerator with parallel-plate passages and the regenerator with a ceramic-disk core. For these regenerators the passages are continuous through the core, so heat conducts from the hot face to the cool face and reduces effectiveness (see Chapter 6). This happens not only when the core is exposed to the flows, but also when the core rotates under the seals. This makes it advantageous to have narrow seals and a small Seal Coverage for these core types. For the Mesh Screen Matrix (MSM), regenerator each layer of screen is in poor thermal contact with the layer above it and the layer below it, so core conduction is insignificant. Optimizations show that Seal Coverages between 10% and 30% are typically optimal (see Section 5.2 for example).

Seal Clearance and Leakage Constant, δ and K_{LEAK}. To minimize leakage, the clearance of the seals that separate the flows should be as small as possible. Thermal distortions and wear can lead to large seal clearances. We have estimated seal clearances of 0.5 mm for the larger applications (20-MW and 3-MW) and 0.2 mm for the 75-kW automotive gas turbine. Harper [26] gives a method for calculating seal leakage in regenerators (see Chapter 6 for a description of Harper's method). Harper [26] shows that a dimensionless total leakage is a function of two dimensionless parameters, BS/A and r_R:

$$\frac{\dot{m}_{Xl\Sigma}}{\dot{m}_{Xlc}} = \frac{\dot{m}_{Xl\Sigma}}{\dot{m}_{Xlc}}\left(BS/A, r_R\right) , \tag{5.2}$$

where

$\dot{m}_{Xlc}$	=	Carry-over leakage (kg/s);
$\dot{m}_{Xl\Sigma}$	=	Total seal leakage, direct and carry-over combined (kg/s);
$\dot{m}_{Xl\Sigma}/\dot{m}_{Xlc}$	=	Dimensionless total seal leakage (–);
BS/A	=	Dimensionless core rotation speed (–); and
r_R	=	Pressure ratio across the seals (–).

Leakage is high if the dimensionless core rotation speed (BS/A) and/or the pressure ratio across the seals (r_R) have large values. The dimensionless core rotation speed is proportional to the product of (dimensional) core speed and a leakage constant, and it is inversely proportional to the seal clearance:

$$\frac{BS}{A} \propto \frac{S_{CORE} K_{LEAK}}{\delta} , \tag{5.3}$$

where

S_{CORE}	=	Core speed (m/s);
K_{LEAK}	=	Leakage constant (–); and
δ	=	Seal clearance (mm).

Therefore leakage is large for fast core rotations and/or large seal clearances. The value for the leakage constant depends on the core surface and the seal surface that bears upon it. A typical value is $K_{LEAK} = 0.6$.

Inner Diameter of the Core, D_i. In Direct Regenerator Design, the designer specifies the inner diameter of the core, D_i. For regenerators in which the compressed air and exhaust flow radially through the core (for example the Mesh Screen Matrix regenerator and the regenerator with parallel-plate passages), the inner core diameter might be selected based on space constraints. In addition, the designer can iteratively specify the inner core diameter (D_i) and calculate the core width (w) to achieve an acceptable aspect ratio for the core. For regenerators in which the flows are parallel to the axis of rotation (for example the regenerator with a ceramic-disk core), the designer can iteratively specify the inner core diameter (D_i) and calculate the outer diameter (D_o).

The remaining regenerator specifications are characterstics of the core.

Porosity, p. Core porosity is defined as the ratio of the voids volume to the total core volume. High core porosity minimizes the total size and weight of the regenerator as well as any reductions in effectiveness due to low porosity (see Chapter 6). However, highly porous cores are light-weight with little heat capacity per unit of core volume, so they must rotate very fast for the temperature profile through the core not to change significantly during each rotation. In other words, highly porous cores rotate quickly to meet a specification for the dimensionless core rotation rate (C_{ROT}). We have selected a core porosity of $p = 0.8$ for the MSM regenerator (20-MW gas-turbine application) and the regenerator with a ceramic-disk core (75-kW gas-turbine application). For the regenerator with parallel-plate passages, we have selected a higher core porosity of $p = 0.85$ to reduce the solid area available for heat conduction from the hot face to the cool face (at the expense of a faster core rotation speed and higher carry-over leakage). Typical porosity values are in the range $0.5 \leq p \leq 0.8$.

Heat-Transfer and Flow-Friction Data, Nu and C_f. We use different Nusselt numbers and friction coefficients for the different regenerator types. All of the passage dimensions are small enough so that the flows are well within the laminar-flow regime (Reynolds numbers are typically about $Re \approx 50$–200). In the laminar-flow regime the Nusselt numbers and friction coefficients have constant values, which Table 5.3 lists. For the Mesh Screen Matrix (MSM) regenerator (for the 20-MW gas turbine) we use data for continuous square passages. This is an approximation. We would expect the heat transfer to be a little better for MSM's and the pressure drops to be a little worse. For the regenerator with parallel-plate passages (for the 3-MW gas turbine) we use data for the parallel-plate geometry (of course), which is the passage geometry that yields the most compact regenerators. For the regenerator with a ceramic-disk core (for the 75-kW gas turbine) we use data for 2×1 rectangular passages, because this geometry is typically used for these core types.

Hydraulic Diameter, D_H. The hydraulic diameter of a passage is defined as four times the ratio of the flow area to the wetted perimeter of the passage. We show below that the total frontal area of a regenerator is independent of the hydraulic diameter, but the core thickness (or flow length) varies as the inverse square of the hydraulic diameter. The cores of current regenerators are about 80 mm thick and have passages with hydraulic diameters of about 0.5 mm. If the hydraulic diameters were reduced to 0.25 mm the cores would need to be only 20 mm thick to have the same convective conductances and pressure drops. Manufacturing and fouling considerations limit how small the passages in ceramic-disk cores can be made. In addition, except for the Mesh Screen Matrix (MSM), thin regenerator cores have high losses due to heat conduction through the core from the hot face to the cool face. The conduction path though an MSM is interrupted between the screen layers, so conduction is insignificant.

Thermal Properties of the Core Material. Table 5.1 lists thermal properties for stainless steel for the Mesh Screen Matrix regenerator and the regenerator with parallel-plate passages (see Mills [50] for these properties). Properties of Cordierite ceramic are listed for the regenerator with a ceramic-disk core (see Kleiner, et al. [39]).

The following section describes the Direct Regenerator Design procedure by describing how to calculate the regenerator dimensions from the designer's specifications.

5.1.2 Five-Step Procedure

Direct Regenerator Design consists of five steps:

1. Calculate gas properties;
2. Calculate regenerator heat-transfer parameters;
3. Calculate core performance;
4. Calculate regenerator dimensions; and
5. Calculate additional outputs.

The following paragraphs describe these steps in detail.

Passage Geometry	Nu	C_f	Nu/C_f	$Nu\,C_f$
	3.00	13.3	0.225	40.0
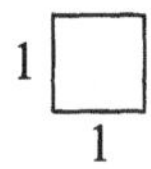	3.61	14.2	0.254	51.3
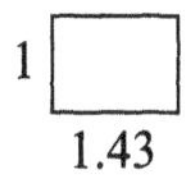	3.73	14.6	0.255	54.5
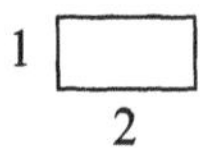	4.12	15.6	0.264	64.3
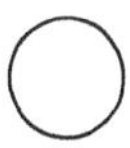	4.36	16.0	0.273	69.8
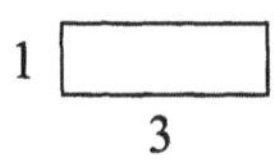	4.79	17.2	0.278	82.4
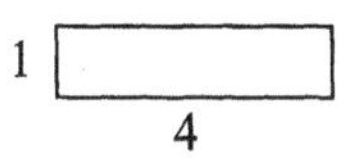	5.33	18.3	0.291	97.5
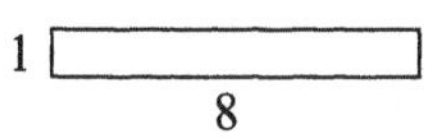	6.49	20.6	0.315	133.7
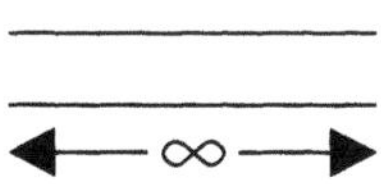	8.235	24.0	0.343	197.6

Table 5.3: Laminar-Flow Heat-Transfer and Pressure-Drop Data for Alternative Passage Geometries (data taken with permission from Kays and Crawford [36] and Kays and London [37]).

Property	Units	Description	20-MW GT	3-MW GT	75-kW GT
T_{No}	K	Outlet Temperature on the Compressed-Air Side	869.2	1,031.5	1,104.1
i_{Ni}	kJ/kg	Inlet Specific Enthalpy on the Compressed-Air Side	460.3	650.7	440.7
i_{No}	kJ/kg	Outlet Specific Enthalpy on the Compressed-Air Side	898.5	1,082	1,166
i_{Xi}	kJ/kg	Inlet Specific Enthalpy on the Exhaust Side	939.8	1,157	1,219
i_{Xo}	kJ/kg	Outlet Specific Enthalpy on the Exhaust Side	514.4	746.4	499.8
T_{Xo}	K	Outlet Temperature on the Exhaust Side	504.7	718.7	493.9
P_{Xo}	MPa	Outlet Pressure on the Exhaust Side	0.1020	0.1020	0.1020
$(\Delta P)_X$	Pa	Pressure Drop Across the Exhaust Side	2,061	2,061	2,061
$\overline{P}_X$	MPa	Average Pressure on the Exhaust Side	0.1031	0.1031	0.1031
$\overline{T}_N$	K	Average Temperature on the Compressed-Air Side	663.8	836.4	771.6
$\overline{T}_X$	K	Average Temperature on the Exhaust Side	697.8	896.8	816.5
$\overline{k}_N$	W/m-K	Average Thermal Conductivity on the Compressed-Air Side	0.05029	0.05960	0.05626
$\overline{k}_X$	W/m-K	Average Thermal Conductivity on the Exhaust Side	0.05218	0.06263	0.05861
$\overline{\mu}_N$	kg/m-s	Average Viscosity on the Compressed-Air Side	3.220E-5	3.727E-5	3.543E-5
$\overline{\mu}_X$	kg/m-s	Average Viscosity on the Exhaust Side	3.325E-5	3.890E-5	3.671E-5
$\overline{\rho}_X$	kg/m^3	Average Density on the Exhaust Side	0.5147	0.4004	0.4398
$\overline{\dot{V}}_X$	m^3/s	Average Volumetric Flow Rate Through the Exhaust Side	95.82	20.04	1.036

Table 5.4: Gas Properties Calculated in Step 1.

Step 1: Calculate Gas Properties. The first step is to calculate various thermodynamic properties of the compressed air and exhaust that flow through the regenerator's core. Table 5.4 lists the results of each calculation for the three applications listed above. We calculate the outlet temperature on the compressed-air side of the core from the effectiveness specification:

$$T_{No} = (1 - \epsilon)T_{Ni} + \epsilon T_{Xi} \ , \tag{5.4}$$

where

T_{No}	=	Outlet temperature on the compressed-air side (K);
ϵ	=	Regenerator effectiveness (–);
T_{Ni}	=	Inlet temperature on the compressed-air side (K); and
T_{Xi}	=	Inlet temperature on the exhaust side (K).

The specific enthalpies at the inlet to each side as well as at the outlet on the compressed-air side are then calculated. We use correlations of the data for compressed air and combustion products given by Chappell and Cockshutt [12] to calculate these specific enthalpies. The outlet specific enthalpy on the exhaust side results from conservation of energy:

$$i_{Xo} = i_{Xi} - \left(\frac{\dot{m}_N}{\dot{m}_X}\right)(i_{No} - i_{Ni}) \ , \tag{5.5}$$

where

i_{Xo}	=	Outlet specific enthalpy on the exhaust side (kJ/kg);
i_{Xi}	=	Inlet specific enthalpy on the exhaust side (kJ/kg);
$\dot{m}_N$	=	Mass flow rate of compressed air (kg/s);
$\dot{m}_X$	=	Mass flow rate of exhaust (kg/s);
i_{No}	=	Outlet specific enthalpy on the compressed-air side (kJ/kg); and
i_{Ni}	=	Inlet specific enthalpy on the compressed-air side (kJ/kg).

We determine the outlet temperature on the exhaust side of the regenerator by iterating with the calculation of specific enthalpy for combustion products until we reach the known outlet specific enthalpy on the exhaust side.

We calculate the pressure drop across the exhaust side from the specified dimensionless pressure drop:

$$P_{Xo} = \left(\frac{2 - (\Delta P/P)_X}{2 + (\Delta P/P)_X}\right) P_{Xi} \ , \tag{5.6}$$

where

P_{Xo}	=	Outlet pressure on the exhaust side (MPa);
$(\Delta P/P)_X$	=	Dimensionless pressure drop across the exhaust side (–); and
P_{Xi}	=	Inlet pressure on the exhaust side (MPa).

Then we calculate the (dimensional) pressure drop across the exhaust side:

$$(\Delta P)_X = P_{Xi} - P_{Xo} \ . \tag{5.7}$$

We use average values for the properties on the compressed-air and exhaust sides to calculate the remaining properties. It would be slightly more accurate to conceptually divide the regenerator's core into multiple sections along the flow paths and to use average properties for each section; however, the slight improvement in accuracy rarely justifies the effort (cost) of doing so. In addition, the values for viscosity and thermal conductivity are rarely known precisely.

We calculate the average density and the volumetric flow rate on the exhaust side as follows[3]:

$$\overline{\rho}_X = \frac{\overline{P}_X}{R\overline{T}_X} \quad \text{and} \tag{5.8}$$

$$\overline{\dot{V}}_X = \frac{\dot{m}_X}{\overline{\rho}_X} \ , \tag{5.9}$$

where

$\overline{\rho}_X$	=	Average density on the exhaust side (kg/m^3);
$\overline{P}_X$	=	Average pressure on the exhaust side (Pa);
R	=	Gas constant for air (J/kg-K);
$\overline{T}_X$	=	Average temperature on the exhaust side (K);
$\overline{\dot{V}}_X$	=	Average volumetric flow rate on the exhaust side (m^3/s);
$\dot{m}_X$	=	Mass-flow rate on the exhaust side (kg/s); and
$\overline{\rho}_X$	=	Average density of the exhaust (kg/m^3).

Step 2: Calculate Regenerator Heat-Transfer Parameters. The goal of this step is to calculate heat-transfer parameters so that we can determine how large our regenerator must be in order to satisfy the effectiveness specification. Numerical procedures have been used to calculate regenerator effectiveness as a function of seven dimensionless parameters (see Beck [7, 6], Bahnke and Howard [2], and Lambertson [40]):

$$\epsilon = \epsilon\left(NTU, C_{RAT}, (hA)', C_{ROT}, \lambda, As', SC\right) \ , \tag{5.10}$$

[3] Fuel-to-air ratios in gas turbines (especially regenerative gas turbines) are typically low, so we can use the gas constant for air instead of the gas constant for combustion products.

Heat-Transfer Parameter	Units	Description	20-MW GT	3-MW GT	75-kW GT
C_N	W/K	Heat-Capacity Rate ($\dot{m}c_P$) on the Compressed-Air Side	51,061	8,451	492.7
C_X	W/K	Heat-Capacity Rate on the Exhaust Side	54,336	9,259	507.8
C_{RAT}	–	Heat-Capacity Rate Ratio	0.9397	0.9127	0.9702
$(hA)'$	–	Convective-Conductance Ratio	0.2892	0.3807	0.2879
As_N	m^2	Solid Area on the Compressed-Air Side	–	0.1265	0.01191
As_X	m^2	Solid Area on the Exhaust Side	–	0.3163	0.03970
λ	–	Conduction Parameter	–	5.309%	0.4102%
As'	–	Solid-Area Ratio	–	0.4	0.3
NTU	–	Number of Transfer Units	14.91	11.91	19.52

Table 5.5: Heat-Transfer Parameters Calculated in Step 2.

where

ϵ	=	Regenerator effectiveness (–);
NTU	=	Number of Transfer Units, a dimensionless heat exchanger size (–);
C_{RAT}	=	Heat-capacity-rate ratio (–);
$(hA)'$	=	Convective-conductance ratio (–);
C_{ROT}	=	Dimensionless core rotation rate (–);
λ	=	Core conduction parameter (–);
As'	=	Solid-area ratio (–); and
SC	=	Seal Coverage, the fraction of the core face area covered by seals (–).

Chapter 1 defines these dimensionless heat-transfer parameters. Beck [4, 6], Shah [61], Coppage and London [14], and Kays and London [37] have developed correlations that accurately match data calculated using the numerical procedures. Chapter 6 summarizes these correlations. Table 5.5 lists results of calculations for our three example applications listed above.

We calculate the heat-capacity rates on each side of the regenerator as follows:

$$C_N \equiv \frac{\dot{m}_N(i_{No} - i_{Ni})}{T_{No} - T_{Ni}} \text{ and} \tag{5.11}$$

$$C_X \equiv \frac{\dot{m}_X(i_{Xo} - i_{Xi})}{T_{Xo} - T_{Xi}} , \tag{5.12}$$

where

C_N = Heat-capacity rate on the compressed-air side (W/K);
C_X = Heat-capacity rate on the exhaust side (W/K);
$\dot{m}$ = Mass flow rate (kg/s);
i = Specific enthalpy (J/kg); and
T = Temperature (K).

The heat-capacity-rate ratio is then calculated as the ratio of the two heat-capacity rates:

$$C_{RAT} \equiv \frac{C_N}{C_X} , \tag{5.13}$$

where

C_{RAT} = Heat-capacity-rate ratio (–).

We calculate the convective-conductance ratio by

$$(hA)' = \frac{\overline{k}_N}{\overline{k}_X} V' , \tag{5.14}$$

where

$(hA)'$ = Convective-conductance ratio (–);
$\overline{k}_N$ = Average thermal conductivity on the compressed-air side (W/m-K);
$\overline{k}_X$ = Average thermal conductivity on the exhaust side (W/m-K); and
V' = Core volume ratio (–), the ratio of the core volume on the compressed-air side to the core volume on the exhaust side.

For Direct Regenerator Design of Mesh Screen Matrix (MSM) regenerators, we have calculated at this point a sufficient number of heat-transfer parameters—ϵ, C_{RAT}, C_{ROT}, and $(hA)'$—to calculate the required Number of Transfer Units (NTU). For Direct Regenerator Design of regenerators with parallel-plate passages and regenerators with ceramic-disk cores (which have significant core conduction) we need to calculate two additional dimensionless heat-transfer parameters (we also need the specified Seal Coverage, SC):

$$\lambda \equiv \frac{k_R(As_N + As_X)}{C_N L} \text{ and} \tag{5.15}$$

$$As' \equiv \frac{As_N}{As_X} , \tag{5.16}$$

where

λ	=	Core conduction parameter (–);
k_R	=	Thermal conductivity of the core material (W/m-K);
As_N	=	Solid area of the core face exposed to the compressed-air flow (m^2);
As_X	=	Solid area of the core face exposed to the exhaust flow (m^2);
C_N	=	Heat-capacity rate of the compressed air (W/K);
L	=	Core thickness, or flow length (m); and
As'	=	Solid-area ratio (–).

The core thickness and the solid areas can be calculated only after the required Number of Transfer Units has been calculated. Therefore, an iterative procedure must be used for Direct Regenerator Design of regenerators with parallel-plate passages and regenerators with ceramic-disk cores.

Step 3: Calculate Core Performance. In this step we calculate two parameters that quantify the performance of the regenerator core:

1. Core Compactness (CC), the amount of potential for heat-transfer (convective conductance) per unit of core volume; and

2. Permeability (κ), which quantifies the pressure-drop characteristics of the core.

We will see in the next step that these are the key parameters that impact the overall dimensions of the regenerator.

Chapter 1 showed that for laminar flow through uniform passages the calculations for Core Compactness and permeability are:

$$CC = \frac{4pNu}{D_H^2} \text{ and} \tag{5.17}$$

$$\kappa = \frac{pD_H^2}{2C_f}, \tag{5.18}$$

where

CC	=	Core Compactness (m^{-2});
κ	=	Permeability (m^2);
p	=	Core porosity, the ratio of voids volume to total volume (–);
Nu	=	Nusselt number (–);
D_H	=	Hydraulic diameter (m); and
C_f	=	Friction coefficient (–).

Table 5.6 lists results for our three example applications.

Performance Parameter	Units	Description	20-MW GT	3-MW GT	75-kW GT
CC	m^{-2}	Core Compactness	1.284E8	1.120E8	5.363E7
κ	m^2	Permeability	2.535E-9	4.427E-9	6.410E-9

Table 5.6: Core Performance Parameters Calculated in Step 3.

Step 4: Calculate Regenerator Dimensions. This step in Direct Regenerator Design calculates the physical dimensions of the regenerator. Table 5.7 lists results for our three example regenerator designs.

We calculate the thickness of the regenerator core and its total frontal area on the exhaust side as follows:

$$L = \left\{ \frac{(\Delta P)_X NTU C_N \left[1 + (hA)'\right]}{\overline{\mu}_X \overline{\dot{V}}_X \overline{k}_X (hA)'} \right\}^{1/2} \left\{ \frac{\kappa}{CC} \right\}^{1/2} \quad \text{and} \tag{5.19}$$

$$A_X = \left\{ \frac{C_N NTU \left[1 + (hA)'\right] \overline{\mu}_X \overline{\dot{V}}_X}{(hA)' \overline{k}_X (\Delta P)_X} \right\}^{1/2} \left\{ \frac{1}{\kappa CC} \right\}^{1/2}, \tag{5.20}$$

where

L	=	Core thickness or flow length (m);
$(\Delta P)_X$	=	Pressure drop across the exhaust side of the core (Pa);
NTU	=	Number of Transfer Units (–);
C_N	=	Heat-capacity rate on the compressed-air side (W/K);
$(hA)'$	=	Convective-conductance ratio (–);
$\overline{\mu}_X$	=	Average viscosity of the exhaust (kg/m-s);
$\overline{\dot{V}}_X$	=	Average volumetric flow rate of the exhaust (m^3/s);
$\overline{k}_X$	=	Average thermal conductivity of the exhaust (W/m-K);
κ	=	Permeability of the core (m^2); and
CC	=	Core Compactness (m^{-2}).

Equations 5.19 and 5.20 show how Core Compactness and permeability impact regenerator dimensions. For a thin core, we need a large ratio of Core Compactness to permeability. For a small frontal area, we need a large product of these two performance parameters. For laminar flow through uniform passages,

$$L \propto \sqrt{\frac{\kappa}{CC}} = \frac{D_H^2}{\sqrt{8 Nu C_f}} \quad \text{and} \tag{5.21}$$

$$A_X \propto \sqrt{\frac{1}{\kappa CC}} = \frac{1}{p}\sqrt{\frac{C_f}{2Nu}}\,. \tag{5.22}$$

Dimension	Units	Description	20-MW GT	3-MW GT	75-kW GT
L	mm	Core Thickness, or Flow Length	28.83	24.68	68.95
A_X	m^2	Total Frontal Area of the Core on the Exhaust Side	17.58	2.109	0.1985
A_N	m^2	Total Frontal Area of the Core on the Compressed-Air Side	5.274	0.8436	0.05955
A_S	m^2	Total Frontal Area of the Core Under Both Seals	2.540	0.3281	0.02867
A_{TOT}	m^2	Total Frontal Area of the Core ($= A_X + A_N + A_S$)	25.40	3.281	0.2867
V_R	m^3	Total Core Volume	0.7322	0.08095	0.01977
m_R	kg	Mass of the Core	1,138	94.35	8.937
$\dot{m}_R$	kg/s	Mass-Flow Rate of the Core Material	333.0	55.11	1.344
τ_{ROT}	s	Rotation Period of the Core	3.417	1.712	6.651
D_o	m	Outer Diameter of the Core	2.058	1.049	0.6063
$\overline{D}$	m	Average Core Diameter	2.029	1.025	0.3281
w	m	Core Width	3.984	1.019	–
w_S	m	Seal Width	0.3187	0.1610	0.05155

Table 5.7: Regenerator Dimensions Calculated in Step 4.

Therefore, for a thin core, we need a small hydraulic diameter and a large product of Nusselt number and friction coefficient. For a small frontal area, we need a highly porous core and a large ratio of Nusselt number to friction coefficient. (Note that the frontal area of the core is independent of hydraulic diameter.) Table 5.3 shows why regenerators with parallel-plate passage geometries are the most compact: both the product of Nusselt number and friction coefficient *and* the ratio of Nusselt number to friction coefficient are higher than for the other geometries.

We calculate some additional core dimensions as follows. Table 5.7 defines these dimensions.

$$A_N = V' A_X \ ; \tag{5.23}$$

$$A_S = \left(\frac{SC}{1-SC}\right)(A_N + A_X) \ ; \tag{5.24}$$

$$A_{TOT} = A_X + A_N + A_S \ ; \tag{5.25}$$

$$V_R = A_{TOT} L \ ; \tag{5.26}$$

$$m_R = \rho_R (1-p) V_R \ ; \tag{5.27}$$

$$\dot{m}_R = \frac{C_{ROT} C_N}{c_R} \ ; \text{ and} \tag{5.28}$$

$$\tau_{ROT} = \frac{m_R}{\dot{m}_R} \ . \tag{5.29}$$

We calculate the outer diameter of the core differently, depending on whether the regenerator has radial flows or axial flows:

for radial-flow regenerators,

$$D_o = D_i + 2L \ ; \text{ and} \tag{5.30}$$

for axial-flow regenerators,

$$D_o = \left(D_i^2 + \frac{4A_{TOT}}{\pi}\right)^{1/2} \ . \tag{5.31}$$

We then calculate the average core diameter for each type of regenerator in the same way:

$$\overline{D} = \frac{D_i + D_o}{2} \ . \tag{5.32}$$

The width of the regenerator core is relevant only for regenerators with radial flows: *for radial-flow regenerators only*;

$$w = \frac{A_{TOT}}{\pi \overline{D}} \ . \tag{5.33}$$

Finally we calculate the width of each face seal:

$$w_S = \left(\frac{SC}{2}\right) \pi \overline{D} \ . \tag{5.34}$$

Step 5: Calculate Additional Outputs. The fifth (and last) step in Direct Regenerator Design is to calculate additional relevant outputs. Table 5.8 defines these outputs and lists results for our three example regenerator designs. We calculate the outlet pressure on the compressed-air side, the average pressure, and the dimensionless pressure drop across the core as follows:

$$P_{No} = \left(P_{Ni}^2 - \frac{2\overline{\mu}_N L \dot{m}_N R \overline{T}_N}{\kappa A_N} \right)^{1/2} ; \tag{5.35}$$

$$\overline{P}_N = \frac{P_{Ni} + P_{No}}{2} \text{ ; and} \tag{5.36}$$

$$\left(\frac{\Delta P}{P} \right)_{Nc} = \frac{P_{Ni} - P_{No}}{\overline{P}_N} . \tag{5.37}$$

Note that for each of our three example applications the dimensionless pressure drop across the core on the compressed-air side is less than the 1–2% required for good flow distribution. Therefore, each regenerator requires a partial screen on the compressed-air side to increase the pressure drop and ensure good flow distribution. The second author advocates this technique.

At this point we need to calculate some parameters so that we can calculate the total seal leakage (direct and carry-over combined) from the compressed-air side to the exhaust side using the method of Harper [26]. We calculate the average density on the compressed-air side and the total volume under both seals as follows:

$$\overline{\rho}_N = \frac{\overline{P}_N}{R\overline{T}_N} \text{ ; and} \tag{5.38}$$

$$V_S = A_S L . \tag{5.39}$$

We calculate the seal length differently, depending on whether the regenerator has radial flows or axial flows:

for radial-flow regenerators,

$$L_S = 2\,(w + L) \text{ ; and} \tag{5.40}$$

for axial-flow regenerators,

$$L_S = D_o - D_i + 2L . \tag{5.41}$$

We calculate an average core rotation speed as follows:

$$S_{CORE} = \frac{\pi \overline{D}}{\tau_{ROT}} . \tag{5.42}$$

Harper's [26] method for calculating total seal leakage requires five parameters:

1. Dimensionless rotation speed of the core from the compressed-air side to the exhaust side, $(BS/A)_+$;

Output	Units	Description	20-MW GT	3-MW GT	75-kW GT
P_{No}	MPa	Outlet Pressure on Compressed-Air Side	1.176	0.9797	0.2919
$\overline{P}_N$	MPa	Average Pressure on the Compressed-Air Side	1.177	0.9800	0.2930
$(\Delta P/P)_{Nc}$	-	Dimensionless Pressure Drop Across the Core on the Compressed-Air Side	0.04572%	0.04703%	0.7459%
$\overline{\rho}_N$	kg/m^3	Average Density of the Compressed Air	6.178	4.083	1.323
V_S	m^3	Total Volume Under Both Seals	0.07322	0.008095	0.001977
L_S	m	Seal Length	8.026	2.088	0.6942
S_{CORE}	m/s	Average Rotational Speed of the Core	1.865	1.880	0.1550
$(BS/A)_+$	-	Dimensionless Rotational Speed of the Core from the Compressed-Air Side to the Exhaust Side	1.353	0.5983	0.1568
$(BS/A)_-$	-	Dimensionless Rotational Speed of the Core from the Exhaust Side to the Compressed-Air Side	-1.256	-0.5779	-0.1524
r_R	-	Pressure Ratio Across the Seals	11.42	9.510	2.843
$(\dot{m}_{XIc})_+$	kg/s	Carry-Over Leakage from the Compressed-Air Side to the Exhaust Side	1.059	0.1641	0.003147
$(\dot{m}_{XIc})_-$	kg/s	Carry-Over Leakage from the Exhaust Side to the Compressed-Air Side	0.08823	0.01610	0.001046
$\dot{m}_{XI\Sigma}$	kg/s	Total Seal Leakage	1.377	0.3030	0.01182
$\dot{m}_{XLI}/\dot{m}_4$	-	Dimensionless Regenerator Inlet Leakage	1.409%	1.943%	1.281%
$\dot{m}_{XLO}/\dot{m}_{13}$	-	Dimensionless Regenerator Outlet Leakage	1.438%	1.983%	1.308%

Table 5.8: Additional Outputs Calculated in Step 5.

2. Dimensionless rotation speed of the core from the exhaust side to the compressed-air side, $(BS/A)_-$;

3. Pressure ratio across the seals, r_R;

4. Carry-over leakage from the compressed-air side to the exhaust side, $(\dot{m}_{XIc})_+$; and

5. Carry-over leakage from the exhaust side to the compressed-air side, $(\dot{m}_{XIc})_-$.

We calculate these five parameters as follows:

$$\left(\frac{BS}{A}\right)_+ = \frac{(V_{SP}/2)\, S_{CORE} K_{LEAK}}{\sqrt{2R\overline{T}_N} w_S D_H L_S \delta} ; \tag{5.43}$$

$$\left(\frac{BS}{A}\right)_- = -\frac{(V_{SP}/2)\, S_{CORE} K_{LEAK}}{\sqrt{2R\overline{T}_X} w_S D_H L_S \delta} ; \tag{5.44}$$

$$r_R = \frac{\overline{P}_N}{\overline{P}_X} ; \tag{5.45}$$

$$(\dot{m}_{XIc})_+ = \frac{V_{RP}\overline{\rho}_N}{\tau_{ROT}} \text{ ; and} \tag{5.46}$$

$$(\dot{m}_{XIc})_- = \frac{V_{RP}\overline{\rho}_X}{\tau_{ROT}} . \tag{5.47}$$

Then we calculate the total seal leakage (direct and carry-over combined) using Harper's [26] method:

$$\begin{aligned} \dot{m}_{XI\Sigma} &= (\dot{m}_{XIc})_+ \left(\frac{\dot{m}_{XI\Sigma}}{\dot{m}_{XIc}}\right)\left([BS/A]_+, r_R\right) + \\ &\quad (\dot{m}_{XIc})_- \left(\frac{\dot{m}_{XI\Sigma}}{\dot{m}_{XIc}}\right)\left([BS/A]_-, r_R\right) . \end{aligned} \tag{5.48}$$

5.2 Optimal Regenerator Design

A typical procedure for designing a regenerative gas turbine is to:

1. Specify various cycle and regenerator parameters (for example pressure ratios and regenerator effectiveness) based on experience;

2. Calculate engine performance; and then

3. Design the regenerator using Direct Regenerator Design.

This procedure is usually carried out iteratively until a high thermal efficiency results as well as an acceptable regenerator cost, size, weight, rotation speed, etc. A better approach is to find optimal values for the cycle and regenerator parameters that maximize thermal efficiency. This is the approach taken in Optimal Regenerator Design.

Optimal Regenerator Design has been included in the commercial software RGT-OPT^{TM} [4]. This section shows how to use Optimal Regenerator Design and RGT-OPT^{TM} to maximize the thermal efficiency of an Intercooled Regenerative (ICR) gas turbine with reheat.

Chapter 3 showed that arbitrary specifications for an ICR cycle with reheat led to a thermal efficiency that was not much better than the thermal efficiency of an otherwise similar ICR cycle without reheat. This was because the goal of Chapter 3 was to illustrate the features of gas-turbine cycles, and no effort was made to optimize the ICR-with-reheat cycle. This section will show that Optimal Regenerator Design enables the designer to design an ICR cycle with reheat that has a maximized thermal efficiency of 52.74%.

In Optimal Regenerator Design for an ICR cycle with reheat we specify most of the same cycle parameters that are listed in Table 3.5. Exceptions are the pressure ratios, which are optimized. In addition, we specify many of the same regenerator parameters as in Direct Regenerator Design (see Table 5.1). Exceptions are regenerator effectiveness (ϵ), dimensionless core rotation rate (C_{ROT}), and Seal Coverage (SC), which are optimized. Table 5.9 lists regenerator specifications for Optimal Regenerator Design.

In Optimal Regenerator Design we must specify minimum and maximum values for six optimization parameters (three cycle pressure ratios and three regenerator parameters):

1. r_{C1} (–), pressure ratio of the *first* compressor;
2. r_{C2} (–), pressure ratio of the *second* compressor;
3. r_{E1} (–), pressure ratio of the *first* turbine;
4. C_{ROT} (–), dimensionless core rotation rate;
5. SC (–), Seal Coverage; and
6. D_o (m), outer diameter of the core.

We list our selected allowable ranges for these parameters as well as the optimal values in Table 5.10. We also list the rest of the optimization results in Table 5.10. Our optimization has resulted in a low overall pressure ratio (about 4.0). The regenerator is relatively large (0.8 m outer diameter for a 100-kW gas turbine), and it therefore has a high effectiveness of 98.72%. We could reduce the size and effectiveness of the regenerator by selecting a lower value for the maximum allowable outer diameter of the core. (Note that the optimization converged on the maximum allowable value.) A reduced outer diameter would make the seal length shorter, so RGT-OPT would be able to increase the pressure ratio to achieve a higher power density and a lower gas flow through the regenerator and still achieve low seal leakage. The smaller flow through the regenerator would boost effectiveness to partially compensate for the reduction caused by the smaller outer diameter of the core.

[4] RGT-OPT is a trademark of Douglas Stephen Beck, the principal author, and is available from him.

Specification	Description	Value
V'	Core Volume Ratio, the Ratio of the Volume on the Compress-Air Side to the Volume on the Exhaust Side	0.3
δ	Seal Clearance	0.2 mm
K_{LEAK}	Seal-Leakage Coefficient	0.6
D_i	Inner Diameter of the Core	0.05 m
p	Core Porosity, the Ratio of the Voids Volume to the Total Volume	0.8
Nu	Nusselt Number	4.91
C_f	Friction Coefficient	15.6
D_H	Hydraulic Diameter	0.5 mm
ρ_R	Density of the Core Material	2,260 kg/m^3
c_R	Specific Heat Capacity of the Core Material	1,100 J/kg-K
k_R	Thermal Conductivity of the Core Material	2.7 W/m-K

Table 5.9: Regenerator Specifications for Optimal Design of an ICR Cycle with Reheat.

Parameter	Description	Min.	Optimal	Max.
Maximum Efficiency:				
η_{TH}	Thermal Efficiency		52.74 %	
Cycle Optimization Parameters:				
r_{C1}	Pressure Ratio of the First Compressor	1.1	2.007	4.0
r_{C2}	Pressure Ratio of the Second Compressor	1.1	1.935	4.0
r_{E1}	Pressure Ratio of the First Turbine	1.1	1.577	4.0
Regenerator Optimization Parameters:				
C_{ROT}	Dimensionless Core Rotation Rate	2.0	4.272	8.0
SC	Seal Coverage, the Fraction of the Core Face Area Covered by Seals	0.01	0.1400	0.8
D_o	Outer Diameter of the Core	0.1 m	0.8 m	0.8 m
Other Parameters:				
ϵ	Regenerator Effectiveness		0.9872	
τ_{ROT}	Core Rotation Period		14.48 s	
L	Core Thickness, or Flow Length		0.1197 m	
m_R	Mass of the Core		27.09 kg	
w_S	Seal Width		0.09349 m	
$\dot{m}_{XLI}/\dot{m}_4$	Dimensionless Regenerator Inlet Leakage		2.020%	
$\dot{m}_{XLO}/\dot{m}_{13}$	Dimensionless Regenerator Outlet Leakage		2.062%	

Table 5.10: Results of Optimal Design of an ICR Cycle with Reheat.

5.3 Method of Kays and London

This section describes a design method attributed to Kays and London [37]. This method is appropriate for preliminary regenerator design, in which the designer seeks estimates of the dimensions of a regenerator that will perform a specified duty. We illustrate this preliminary-design method with two examples. Section 5.3.1 considers a design of a ceramic-disk rotary regenerator for a 1 MW regenerative gas turbine. Section 5.3.2 describes a modular-regenerator design for a 10 MW regenerative gas turbine.

5.3.1 Rotary Regenerator for a 1 MW Gas Turbine

We consider the preliminary design of a regenerator for a 1-MW gas turbine running on the simple regenerative cycle. We begin under the assumption that the designer of the overall machine has given the desired (specified) effectiveness and/or the inlet and outlet temperatures, compositions, and mass-flow rates of both flows, guidelines on the two normalized pressure drops, and the following specifications:

Compressor mass-flow rate, $\dot{m}_C =$	3.3 kg/s;
Turbine mass-flow rate =	3.35 kg/s;
Compressor-outlet temperature =	468 K;
Compressor-outlet pressure =	4.25 bar;
Turbine-outlet temperature =	1,350 K;
Effectiveness	95%; and
Exhaust-side pressure drop	$\leq$ 2%.

It is hoped that the designer of the overall machine has recognized that the regenerator will be the backbone of her or his system and has involved the regenerator designer early in the machine design process. After the regenerator has been designed, the designer of the overall machine will probably want to modify her or his effectiveness specification, raising or lowering it based on the dimensions of the regenerator design.

The steps in this design follow:

1. Figure 6.5 was generated from data calculated using the methods of Lambertson [40], Bahnke and Howard [2], and Beck [5], and it shows the heat-transfer performance of regenerators with equal heat-capacity rates ($C_{RAT} = 1$) and no conduction ($\lambda = 0$). An approximation of equal heat-capacity rates is usually conservative for design of gas-turbine regenerators because the heat-capacity rate of the compressed-air flow is usually less than that of the exhaust flow, and regenerator effectiveness is greatest when the heat-capacity rate of the compressed-air flow is less than that of the exhaust flow. Thus, a gas-turbine regenerator designed in part under the approximation of equal heat-capacity rates will probably be oversized and will probably achieve an effectiveness slightly higher than the specified effectiveness.

We enter Figure 6.5 with the specified effectiveness of 95% and obtain the required value for the Number of Transfer Units: $NTU = 22$. The Number of Transfer Units is

$$NTU \equiv \left[\frac{1}{\dot{m}_N c_{PN}}\right]\left[\frac{1}{1/(hA)_N + 1/(hA)_X}\right] . \tag{5.49}$$

The product in the denominator of the first term is the minimum (indicated with an "N") heat-capacity rate (product of mass-flow rate and specific heat capacity). As stated above, for most gas-turbine regenerators this will be the heat-capacity rate of the compressed-air flow. The second term is the overall conductance of the regenerator.

2. We can find the mean properties of the two fluids from the above data and from property data such as that from the National Bureau of Standards and elsewhere given in Wilson [73]. We are performing this preliminary-design calculation using what may seem to be gross approximations: the mean properties of the two fluids. However, such calculations are fully adequate for preliminary design. It is usual to use the products of a preliminary-design calculation in a computer program that divides the regenerator into many "slices" and performs an energy balance on each slice using mean properties for each slice. Such a calculation procedure is quite accurate, and the results are usually quite close to those from the preliminary design.

   Figure 5.4 shows the temperatures through the regenerator. We obtain the compressed-air outlet temperature from the specified effectiveness, and the exhaust outlet temperature from an overall energy balance on the regenerator using successive iterations to find the mean specific heat capacity of the exhaust. We obtain the following data:

	Compressed Air ("N")	Exhaust ("X")
T_i, K	468	1,350
T_o, K	1,305.9	533.2
$\overline{T}$, K	887	941.6
$\overline{c}_p$, J/kg-K	1,117.6	1,129.5
$\overline{\mu}$, kg/m-s	3.864E-5	4.002E-5
$\overline{P}$, bar	4.25	1.01
$\overline{\rho}$, kg/m^3	1.670	0.3737
Pr	0.695	0.698
$Pr^{1/3}$	0.886	0.887
k, W/m-K	6.214×10^{-2}	6.484×10^{-2}
h, W/m^2-K	323.5	338.0

   From these data we obtain

$$\frac{1}{1/(hA)_N + 1/(hA)_X} = 22 \times 3.3 \times 1,117.7 = 81,140 \text{ W/K}. \tag{5.50}$$

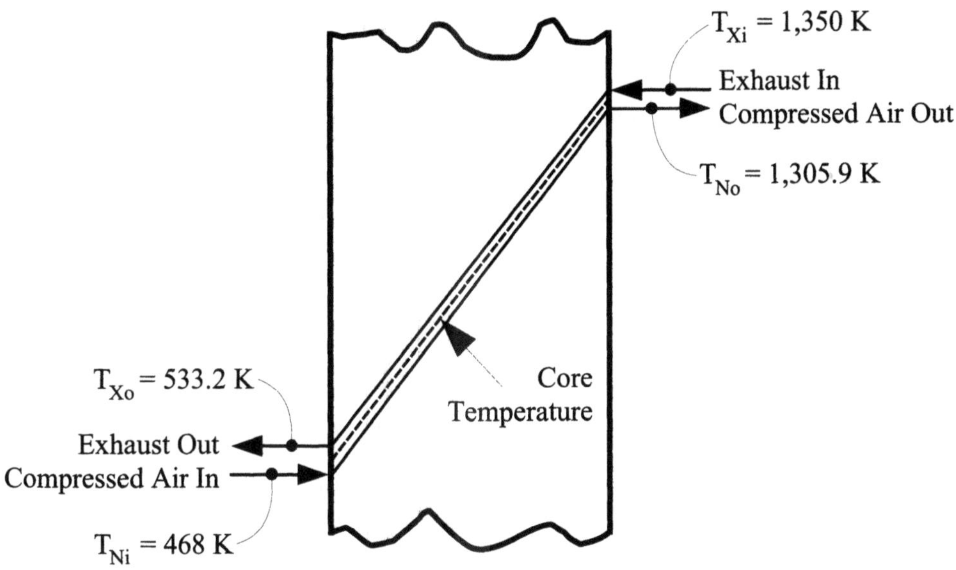

Figure 5.4: Regenerator Temperatures.

3. This regenerator-design method gives the designer some freedom at this point: she or he must specify the convective-conductance ratio, $(hA)' \equiv (hA)_N/(hA)_X$. She or he can decide to use equal convective conductances $[(hA)' = 1]$. This choice will yield a minimum core volume, but will produce highly unequal core pressure drops: large on the exhaust side and small on the compressed-air side. Neither is desirable. A large pressure drop on the exhaust side requires a large pumping-power expenditure and reduces thermal efficiency and specific power. A small pressure drop on the compressed-air side makes the flow distribution very sensitive to the header design, and flow maldistribution has a negative effect on regenerator effectiveness (see Wilson [72]; London, et al. [44]; London [43]; and Section 1.2.9): a regenerator with flow maldistribution must be made larger to achieve a specified effectiveness. To avoid the highly unequal pressure drops, the designer could choose a convective-conductance ratio $(hA)' < 1$. We would like to facilitate the possible later choice of the use of a discontinuously rotating regenerator, for which it is beneficial to split the face area into sectors that allow the rotation increments to be a simple fraction of 360°. We will choose a rotational increment of $\alpha_{ROT} = 45^o$ (see Figure 1.12) and a consequent compressed-air-side sector of 135° $(\Delta = 45^o)$ and an exhaust-side sector of 225°. This gives a face-area ratio (compressed-air-to-exhaust) of 5/3 (which should be adjusted in a second design iteration to allow for the width of the radial seals).

4. The next step is to choose a core configuration. We choose a Corning regenerator core designated Cercor L689, the passages of which have hy-

draulic diameters $D_H = 0.5105$ mm. The porosity of the core is $p = 0.708$. Other characteristics of the core will be introduced as needed.

5. One characteristic common to all cores with small hydraulic diameters is that the heat-transfer coefficient is unaffected by flow velocity. This is a consequence of there being laminar flow in the channels because of the low Reynolds numbers. Thus we can calculate the mean heat-transfer coefficients at this point by using a constant Nusselt number (dimensionless heat-transfer coefficient):

$$Nu \equiv \frac{hD_H}{k} = 3.0Pr^{1/3}. \tag{5.51}$$

   Values for the mean heat-transfer coefficients, $\overline{h}$, are given above.

6. Given the heat-transfer coefficients and the ratio between the two areas, Equation 5.50 is soluble. We then have the heat-transfer areas of the two sides:

$$\frac{1}{81,140} = \frac{1}{323.5A_{hN}} + \frac{1}{(828.0)(5/3)(A_{hN})}. \tag{5.52}$$

   Therefore,

$$\begin{aligned} A_{hN} &= 250.8 + 144.0 = 394.8 \text{ m}^2\text{; and} \\ A_{hX} &= (5/3)A_{hN} = 658.1 \text{ m}^2. \end{aligned} \tag{5.53}$$

7. Another characteristic of the core configuration is that it has a specified value of the ratio of the heat-transfer area to the core volume: $A_h/V_R = 5,551 \text{ m}^2/\text{m}^3$. Thus the active volume of the two sides of the regenerator can be found:

$$\begin{aligned} V_N &= \frac{394.8}{5,551} = 0.0711 \text{ m}^3; \\ V_X &= \frac{658.1}{5,551} = 0.1186 \text{ m}^3\text{; and} \\ \sum V &= 0.1897 \text{ m}^3. \end{aligned} \tag{5.54}$$

8. In this design method, we arrive at the face areas of the heat exchanger by choosing one of the mean velocities (or Reynolds number or Mach number, which amount to the same choice). For gas-turbine regenerators the mean velocities normally fall into the range $3 \le \overline{U} \le 30$ m/s. Based on previous experience we make a first choice of $\overline{U}_X = 6$ m/s for the exhaust-side mean velocity. This enables us to find the total cross-sectional area of the flow channels on each side, which we call the free-face area (A_{ff}). If we divide this area by the porosity, we get the actual face area, A_f. From these areas and the regenerator's active volume (the core volume under

the seals, V_S, is inactive) we find the core thickness in the flow direction, L:

$$\begin{aligned}
A_{ffX} &= \frac{\dot{m}_X}{\overline{\rho}_X U_X} = \frac{3.35}{0.3737 \times 6} = 1.494 \text{ m}^2; \\
A_{fX} &= \frac{A_{ffX}}{p} = \frac{1.494}{0.708} = 2.110 \text{ m}^2; \\
A_{fN} &= 2.110 \times 3/5 = 1.266 \text{ m}^2; \\
\sum A_f &= 3.376 \text{ m}^2; \text{ and} \\
L &= \frac{\sum V}{\sum A_f} = \frac{0.1897}{3.376} = 56.2 \text{ mm.} \qquad (5.55)
\end{aligned}$$

(5.56)

9. We find the normalized pressure drops on each side, using the simple relation for the (constant) dimensionless friction coefficient in laminar flow:

$$C_f \equiv f Re = \left[\frac{\tau_{wall}}{(1/2)\rho U^2}\right]\left[\frac{\rho U D_H}{\mu}\right] = \frac{(\Delta P)_{CORE} D_H^2}{2LU\mu} = 14.0. \qquad (5.57)$$

This leads to

$$(\Delta P)_X = 241.7 \text{ Pa, and} \qquad (5.58)$$

$$\left(\frac{\Delta P}{P}\right)_X = \frac{241.7}{101{,}000} = 0.24\%. \qquad (5.59)$$

This pressure drop is too low for good flow distribution, so we choose a higher value for the mean exhaust-side velocity in a second iteration: $\overline{U}_X = 12$ m/s. The free face area of both sides is halved, the flow length is doubled, and the new exhaust-side core pressure drop is 0.96%.

The pressure drop on the compressed-air side is calculated in a similar fashion:

$$\begin{aligned}
\overline{U}_N &= \frac{\dot{m}_N}{\overline{\rho}_N A_{ffN}} = \frac{3.3}{1.670 \times 0.7474 \times 3/5} = 4.410 \text{ m/s;} \\
(\Delta P)_N &= 7 \times \frac{3.864}{10^5} \times \frac{1{,}000}{0.5105}\left(\frac{394.8}{1.266 \times 1/2}\right) = 1{,}456 \text{ Pa; and} \\
\left(\frac{\Delta P}{P}\right)_N &= \frac{1{,}456}{4.25 \times 10^5} = 0.34\%. \qquad (5.60)
\end{aligned}$$

This pressure drop is too small to give good flow distribution, so flow distributors must be used.

10. As stated above, the pressure drops are dependent variables. However, once this stage has been reached, one can adjust the mean velocity through one side of the regenerator to raise or lower the pressure drop on that side, without changing the thermal performance. Therefore, one pressure drop can be regarded as an independent variable, at least for laminar-flow heat exchangers.

11. The carryover leakage is the total mass flow of the compressed air trapped in the regenerator passages as they pass under the radial seals into the exhaust flow. We need to calculate the rotation rate of the ceramic core. This rate is dependent on the dimensionless rotation rate chosen above: $C_{ROT} = 3$. This means that the product of the mass-flow rate of the ceramic and the mean specific heat of the ceramic material must be 3 times the minimum heat-capacity rate:

$$\begin{aligned} C_R &= (\dot{m}\bar{c})_{CERAMIC} = 3.0 C_N = 3.0\,(\dot{m}\bar{c}_P)_N \\ &= 3.0 \times 3.3 \times 1,117.6 \text{ W/K}; \\ \dot{m}_{CERAMIC} &= 9.621 \text{ kg/s; and} \\ \rho_{CERAMIC} &\approx 2,260 \text{ kg/m}^3. \end{aligned} \tag{5.61}$$

Therefore, the ceramic solid flow is

$$\dot{V}_R = \frac{9.621}{2,260} = 0.00426 \text{ m}^3/\text{s}; \tag{5.62}$$

the flow of trapped compressed air is

$$\dot{m}_L = \left(\frac{0.708}{1 - 0.708}\right) \times 0.00426 \times 1.670 = 0.0172 \text{ kg/s}; \tag{5.63}$$

and the leakage is

$$\frac{\dot{m}_L}{\dot{m}_C} = 0.52\%. \tag{5.64}$$

12. The approximate dimensions of the disks, if we specify that there are two, can be found from the face area. We will choose to use hollow-center disks with inside-to-outside diameter ratios of 0.5, and to estimate that the total face area will be increased by 10% to allow for radial seals. (There will also be solid rims on the inside and outside diameters, also with seals, but we are not concerned with these at this stage.) For two regenerator disks,

$$\begin{aligned} \frac{\pi D_o^2}{4}\left(1 - 0.5^2\right) &= \frac{3.376 \times 1.1}{2 \times 2}; \\ D_o &= 1.26 \text{ m}; \\ D_i &= 630 \text{ mm; and} \\ L &= 112.4 \text{ mm}. \end{aligned} \tag{5.65}$$

13. Iterating to improve the regenerator design at this point is very simple. Most of the calculated values do not change, and one can investigate a large number of alternative configurations. The principal changes will be to one of the velocities and to the area-ratio or conductance-ratio split, but changes to the core can also be made rapidly.

5.3.2 Modular Regenerator for a 10 MW Gas Turbine

To simplify the calculations for this example, we will specify that it is a regenerator for a 10-MW gas turbine, having 10 times the mass-flow rates and convective conductances of the previous example. Otherwise, all temperatures, pressures, and other properties and heat-transfer rates are identical.

The point of departure from the previous example is in the choice of convective-conductance ratio. The simplest mechanical arrangement for the modular regenerator is to have equal-area heat-exchanging faces for the cold and hot sides. The heat-transfer areas on each side of the regenerator are then equal as well:

$$A_{hN} = A_{hX} = 811,380\left[\frac{1}{323.5} + \frac{1}{338}\right] = 2,508 + 2,400 = 4,908 \text{ m}^2. \quad (5.66)$$

The total active heat-transfer area is then

$$\sum A_h = 9,817 \text{ m}^2. \quad (5.67)$$

The total active volume of the regenerator core is

$$\sum V_R = \frac{9,817}{5,551} = 1.768 \text{ m}^3. \quad (5.68)$$

We choose a mean exhaust-side velocity of, again, $\overline{U}_X = 12$ m/s. The free-face area for the exhaust side (and the compressed-air side) is given by

$$A_{ffX} = \frac{\dot{m}_X}{\overline{\rho}_X \overline{U}_X} = \frac{33.5}{0.3737 \times 12} = 7.470 \text{ m}^2. \quad (5.69)$$

The porosity is used to calculate the face area:

$$A_{fX} = \frac{7.470}{0.708} = 10.55 \text{ m}^2. \quad (5.70)$$

This and the active volume give the flow length (or core thickness):

$$L = \frac{1}{10.55}\frac{0.884}{2} = 83.8 \text{ mm}. \quad (5.71)$$

We use the same equation as in the first example to calculate the pressure drop on the exhaust side:

$$\begin{aligned} (\Delta P)_X &= 7 \times \frac{4.002}{10^5} \times \frac{10^3}{0.5105} \times 12 \times \frac{4,909}{7.47} = 4,327 \text{ Pa}. \\ \left(\frac{\Delta P}{P}\right)_X &= 4.3\%. \end{aligned} \quad (5.72)$$

This value is higher than for the first example, and could be lowered by reducing the exhaust-side mean velocity. With equal face areas for the exhaust and compressed-air sides, the compressed-air-side pressure drop will be quite low, and distributors will be required to give good flow distribution.

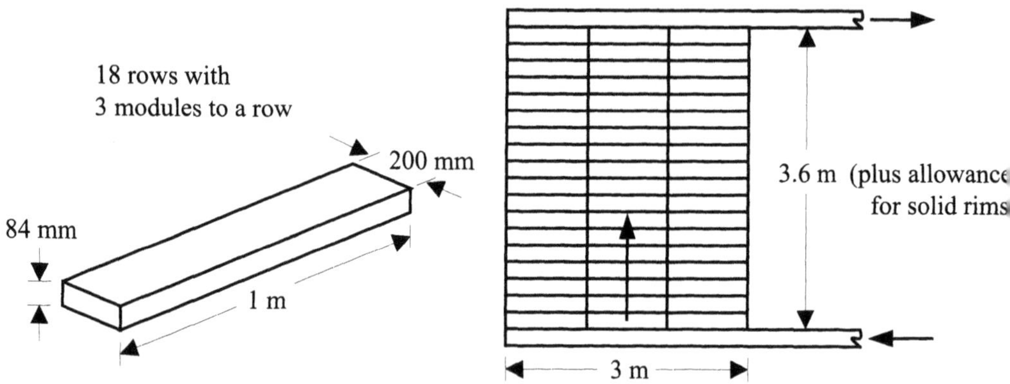

Figure 5.5: Modular-Regenerator Configuration.

The carry-over leakage will be identical to the first case. The ceramic-flow rate will be ten times that of the first example: $\dot{m}_R = 96.21$ kg/s. We choose modules that are 1-m long, 200-mm wide, and 84-mm thick. The active mass of each module (for a ceramic density of 2,260 kg/m^3) is

$$1 \times 0.2 \times 0.084 \times 2,260 = 38 \text{ kg}. \tag{5.73}$$

We could move three modules every 1.2 s to achieve the desired rate. Each heat-exchanger face could be composed of eighteen rows of three-row modules, as Figure 5.5 shows.

5.3.3 Introduction of Hot-Side Pressure Drop, Matrix Temperature Gradient, Cell Shape and Hydraulic Diameter as Design Parameters

As we have indicated above, heat-exchanger design using the method of Kays and London [37] is iterative. The designer chooses inputs, calculates the dimensions, velocities, pressure drops and so forth in the resulting regenerator, and makes changes to bring the design closer to an optimum. We can shorten this process by treating some of the most significant outputs as inputs.

A limiting output in a gas-turbine regenerator is the hot-side pressure drop. In the beginning designer's first attempts at preliminary design, this pressure drop is usually too large (we would characterize a value of $\Delta P/P$ over ten percent as being too large), while the cold-side (compressed-air) pressure drop is often too small (e.g. $\Delta P/P$ under 0.5 percent).

Specification of Hot-Side Pressure Drop

The hot-side core pressure drop is the more critical. We shall show that, once the initial choices of nondimensional rotation rate C_{ROT} and conductance ratio $(hA)'$ have been made, this pressure drop is determined solely by the hot-side mean flow velocity in the core, regardless of the length, shape, or hydraulic diameter of the flow passages, and therefore may be specified as an input.

We develop the approach from first principles by considering a force balance between the pressure drop along a cell passage and the skin-friction on the cell walls:

$$(\Delta P)A = \tau_w A_h \ , \tag{5.74}$$

where

A	$\equiv$	Passage cross-sectional area (m^2);
τ_w	$\equiv$	Skin-friction shear stress (Pa); and
A_h	$\equiv$	Cell-wall area for heat transfer (m^2).

This can be converted to dimensionless groups as follows.

$$\left(\frac{\Delta P}{P}\right)\left[\frac{\tau_w}{(1/2)\rho U^2}\right]\left[\frac{A_h}{A}\right]\left[\frac{(1/2)\rho U^2}{P}\right] \ , \tag{5.75}$$

where

U	$\equiv$	Mean flow velocity through the passage (m/s);
$\tau_w/[(1/2)\rho U^2]$	$\equiv$	f, the friction factor (–); and
ρ	$\equiv$	Mean density of the fluid (kg/m^3).

Therefore, our force balance yields:

$$\left(\frac{\Delta P}{P}\right)_X = f_X \left(\frac{A_{hX}}{A_X}\right) \frac{(1/2)\rho_X U_X^2}{P_X} \ . \tag{5.76}$$

The continuity equation gives

$$\dot{m}_X = \rho_X A_X U_X \ , \tag{5.77}$$

where

$\dot{m}_X$	$\equiv$	Mass flow rate through the passage (kg/s).

Reynolds' analogy relates heat transfer, in the form of the Stanton number, to the skin friction:

$$\text{Stanton no.} \equiv \frac{h_X}{\rho_X U_X c_{PX}} = \frac{1}{2} f_X \ . \tag{5.78}$$

where

h_X	$\equiv$	Heat-transfer coefficient on the hot side (W/m^2-K); and
c_{PX}	$\equiv$	Specific heat capacity at constant pressure on the hot side (J/kg-K).

When this is brought into the force balance, we eventually obtain:

$$\left(\frac{\Delta P}{P}\right)_X = \left[\frac{(hA)_X}{\dot{m}_X c_{PX}}\right]\left[\frac{U_X^2}{RT_X}\right] \ , \tag{5.79}$$

where

$R \equiv$ Gas constant for air or combustion products (J/kg-K); and
$T_X \equiv$ Mean temperature on the hot side (K).

Here the velocity relation is a Mach-number function, and the relation that includes the hot-side conductance is a form of hot-side NTU (Number of Transfer Units).

We remind readers that the hot-side conductance is obtained during preliminary design as follows. The designer starts with the specified input and output temperatures, mass flows, and fluid properties for the regenerator. The designer chooses a value of C_{ROT}, the nondimensional rotation rate (we recommend an initial choice of 3) in order to obtain a value for the NTU required, from curves like the ones given by Kays and London [37] (see Figures 6.5–6.7, which are plots of effectiveness data from Bahnke and Howard [2]). We calculate the overall conductance and the hot-side and cold-side conductances from the definition of NTU (Equation 5.49):

$$\frac{1}{(hA)_N} + \frac{1}{(hA)_X} = \frac{1}{(\dot{m}_N c_{PN})NTU} . \tag{5.80}$$

The designer then chooses a conductance ratio, $(hA)'$. We recommend a starting value of one-third. The hot-side conductance is then given by:

$$(hA)_X = \left[\frac{1}{(hA)'} + 1\right] [(\dot{m}_N c_{PN})NTU] . \tag{5.81}$$

The required hot-side mean velocity to produce the specified core pressure drop is found from:

$$U_X^2 = RT_X \left[\frac{\dot{m}_X c_{PX}}{(hA)_X}\right] \left[\frac{\Delta P}{P}\right]_X . \tag{5.82}$$

As stated above, this velocity has been determined without choosing any details of the core size or shape.

As an example, let us find the mean hot-side velocity required for the regenerator specified by Figure 5.6. We choose $C_{ROT} = 3$ and the conductance ratio $(hA)' = 1/3$. We choose an effectiveness of $\epsilon = 365/428 = 0.853$. The required Number of Transfer Units is then $NTU \approx 8.0$ (see Figures 6.5 and 6.7 in Chapter 6). From Equation 5.81,

$$(hA)_X = (3 + 1)(0.488 \times 1,056.5 \times 8.0) = 16,498 \; W/K . \tag{5.83}$$

We will try a design for a hot-side core pressure drop of 2% [$(\Delta P/P)_X = 0.02$]. From Equation 5.82,

$$\begin{aligned} U_X^2 &= \left(\frac{0.489 \times 1,084.3}{16,498}\right) 0.02 \times 286.96 \times 689 \text{ , or} \\ U_X &= 11.27 \text{ m/s.} \end{aligned}$$

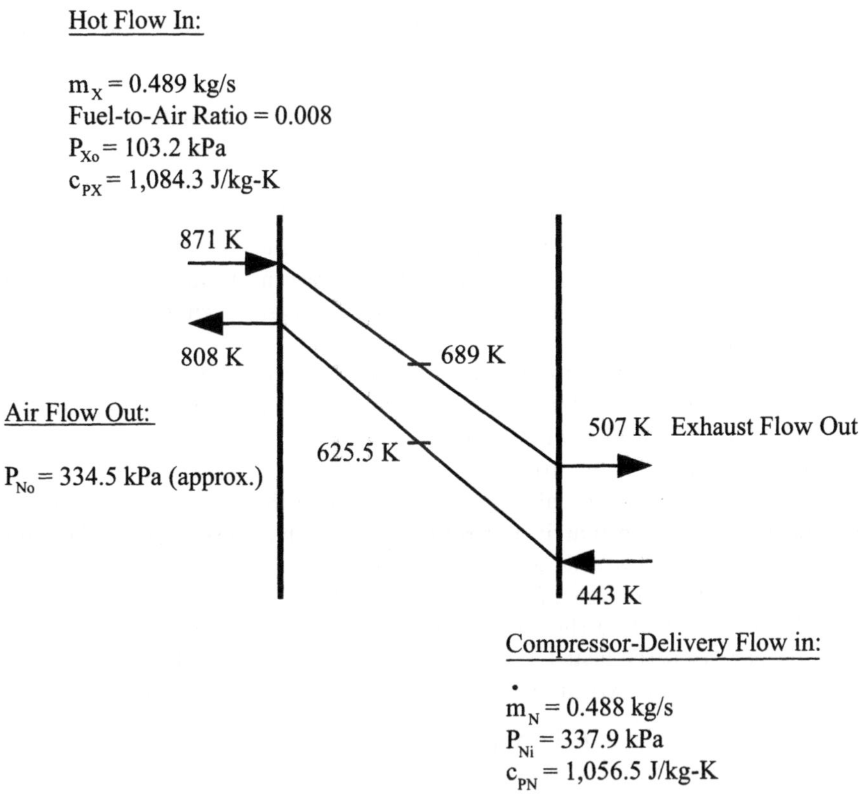

Figure 5.6: Specifications for Design Example.

Specification of Maximum Temperature Gradient

We next consider the temperature gradient in the core. When the core has reached steady state, one face is at a temperature between that of the compressed-air inlet and the exhaust outlet, while the other face is between the turbine-exhaust temperature and the compressed-air outlet temperature. The entire cell-wall temperatures fluctuate a little as the cells pass from the turbine-exhaust flow under the seals to the compressor-delivery flow. The difference between the mean face temperatures is of the order of 500 K. Although the materials used for ceramic cores generally have low thermal-expansion coefficients, temperature differences of this magnitude in a monolithic structure cause stresses, and cores experience failure through cracking. These failures probably result from low-cycle fatigue from the major stress changes during engine start-up and shut-down, but it is possible that the higher-frequency smaller temperature fluctuations during core rotation might also have some added effect. In any case it is desirable to limit the steady-state temperature gradient in the matrix. We suggest using the compressed-air temperature change as a measure of the temperature gradient in the matrix core, (it is a slightly greater gradient than the mean cell-wall gradient) and limiting it to between five and ten K per mm. This will then fix a minimum value for the cell-passage length, which is the matrix or core thickness L.

In the above example, the compressed air undergoes a temperature rise of 365 K. Choosing to use a limiting temperature gradient of 5 K/mm leads to a minimum core thickness of about 75 mm.

The active open volume (or voids volume) for the regenerator core has now been determined. (The actual volume must include the cell-wall material, usually specified from the matrix porosity. This can be chosen later.) We have calculated the free-flow area for the hot side, A_{ffX}. This will be in the same relation to the free-flow area for the cold side, A_{ffN} , as the heat-transfer area ratio A_{hX}/A_{hN}. We find the latter from the conductance ratio $(hA)'$ and from the knowledge that the heat-transfer coefficients for the two sides have the same Nusselt numbers and hydraulic diameters (neither of which is yet chosen).

$$\left(\frac{A_{ffN}}{A_{ffX}}\right) = \left(\frac{A_{hN}}{A_{hX}}\right) ; \tag{5.84}$$

$$(hA)' \equiv \frac{(hA)_N}{(hA)_X} ; \text{ and}$$

$$\frac{h_N}{h_X} = \frac{k_N}{k_X} .$$

Therefore, the free-flow area on the cold side is

$$A_{ffN} = A_{ffX}\frac{k_X}{k_N}(hA)' . \tag{5.85}$$

The active open volume for the regenerator is the sum of the free-flow areas multiplied by the length:

$$\text{Open volume} = (A_{ffX} + A_{ffN})L . \tag{5.86}$$

Choice of Size and Shape of Cells

In Direct Regenerator Design, we specified the hydraulic diameter of the core passages and saw that the total frontal area of a regenerator core is independent of the hydraulic diameter of the passages, but the thickness of the core is proportional to the square of hydraulic diameter (see Section 5.1). In the extension to the design method of Kays and London [37] described in this section, hydraulic diameter is a calculated output. In the following, we show how to calculate the required hydraulic diameter to satisfy specifications, and how it is influenced by cell shape.

Early gas-turbine regenerators were made of alternate strips of wavy and flat stainless steel, wound around a mandrel. Early ceramic cores were similarly constructed, except that paper ribbon saturated in ceramic slurry was used. The consequent cell shapes were approximately triangles with a curved apex and cusped corners, and, as would be expected, were not uniform. Later research showed that lack of cell uniformity can significantly reduce effectiveness. The development of extruded cores allowed a variety of other cross-sections to be produced to great uniformity.

Data from Kays and London [37] and Kays and Crawford [36] show that each passage cross-section has a specific non-dimensional heat-transfer coefficient, the Nusselt number, ranging from 3 for a triangular cross-section to over 8 for parallel plates (see Table 5.3). It seems desirable at first sight to choose to use the cross-section with the highest value of the Nusselt number, Nu. However, the actual heat transferred is determined by the product of the actual heat-transfer coefficient, which is proportional to the Nusselt number divided by the hydraulic diameter, Nu/D_H, and the heat-transfer area. Parallel plates, having no connecting walls, would seem to have a smaller area/volume ratio than, for instance, square passages. The following development shows that in many cases we have considerable freedom to choose among several combinations of cross-section shape and hydraulic diameter.

The hydraulic diameter of a passage is defined as

$$D_H \equiv \frac{4A}{\wp} , \tag{5.87}$$

where

$A \equiv$ Cross-sectional area (m^2); and
$\wp \equiv$ Wetted perimeter (m).

The number of cell passages in the hot face is A_{ffX}/A. Therefore the heat-transfer area is

$$A_{hX} = \left(\frac{A_{ffX}}{A}\right) \wp L . \tag{5.88}$$

The mean heat-transfer coefficient comes from the definition of Nusselt number, $Nu \equiv hD_H/k$:

$$h = \frac{kNu}{D_H} , \tag{5.89}$$

where

$k \equiv$ Thermal conductivity of the fluid (W/m-K).

From these we can derive an expression for the hydraulic diameter required to satisfy the design specification:

$$(hA)_X = \left(\frac{k_X Nu}{D_H}\right) A_{ffX} \left(\frac{\wp}{A}\right) L = \frac{4k_X Nu A_{ffX} L}{D_H^2} . \tag{5.90}$$

Therefore, the passage hydraulic diameter is

$$D_H = \sqrt{\frac{4k_X Nu A_{ffX} L}{(hA)_X}} . \tag{5.91}$$

The use of this is best illustrated by continuing the design example. From the continuity equation, $\dot{m}_X = \rho_X U_X A_{ffX}$. We calculate the hot-side density from the ideal gas law:

$$\rho_X = \frac{102,450}{286.96 \times 689} = 0.5182 \text{ kg/m}^3, \tag{5.92}$$

where we have used the gas constant for air, R=286.96 $J/kg - K$. The free-flow area on the hot side is then

$$A_{ffX} = 0.0837 \text{ m}^2. \tag{5.93}$$

We will also need the mean hot-side thermal conductivity, k_X, which from tables is 0.05169 W/m-K. Table 5.11 lists the hydraulic diameter and wall spacing for alternative passage geometries. The optimum choice is not obvious without cost estimates from the matrix-cell manufacturer. The last three cases are interesting, showing equal heat-transfer performance from cells having almost identical wall spacing with and without connecting walls. The porosity, or cell-wall thickness, can be chosen independently.

5.4 Summary

This chapter has described three alternative methods for regenerator design:

1. Direct Regenerator Design;
2. Optimal Regenerator Design; and
3. The design method of Kays and London [37].

Direct Regenerator Design is used to design regenerators directly from the designer's specifications. Optimal Regenerator Design is used to find optimal regenerator dimensions as well as optimal cycle pressure ratios that maximize cycle

Passage Geometry	b/a	Nu	D_H, mm	a, mm
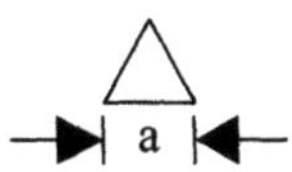	—	3.00	0.486	0.841
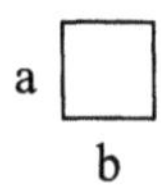	1	3.61	0.533	0.533
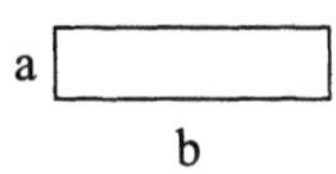	4	5.33	0.648	0.405
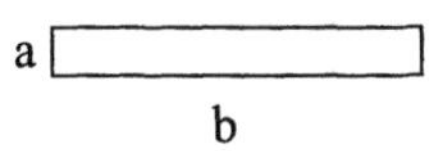	8	6.49	0.715	0.402
	∞	8.235	0.805	0.402

Table 5.11: Hydraulic Diameters and Wall Spacings for Alternative Passage Geometries. Data taken with permission from Kays and Crawford [36]

thermal efficiency. The method of Kays and London [37] is appropriate for preliminary design calculations. An example optimization of an Intercooled Regenerative (ICR) cycle with reheat using the commercial software RGT-OPT™ [5] shows that thermal efficiencies of over 50% should be achievable for automotive gas turbines with ceramic components.

[5]RGT-OPT is a trademark of Douglas Stephen Beck, the principal author, and is available from him.

Chapter 6

Regenerator Performance

This chapter quantifies the performance of gas-turbine regenerators. Chapter 5 showed how to design alternative regenerators for high performance. A key step in all regenerator design procedures is to calculate how a regenerator will perform in terms of heat-transfer, leakage, pressure drops, and power consumption. This chapter shows how to calculate these performances. Section 6.1 covers heat transfer, and Sections 6.3 and 6.4 discuss leakage and pressure drops, respectively. Power consumption is highly dependent on the particular regenerator design and hardware used, so we do not discuss it in detail.

6.1 Heat Transfer

This section describes heat transfer in gas-turbine regenerators. We begin by summarizing how to calculate regenerator effectiveness in Section 6.1.1. Then the rest of this section looks at heat transfer in detail.

6.1.1 Effectiveness Correlations

The rest of this chapter describes how to calculate with a high degree of accuracy the effectiveness of regenerators using sophisticated numerical procedures. Here, we summarize the key results that are relevant for preliminary design work using Direct Regenerator Design, Optimal Regenerator Design, or the design method of Kays and London [37] (see Chapter 5).

Under ideal conditions, the effectiveness of a regenerator approaches the effectiveness of an ideal counterflow recuperator:

$$\epsilon_{CF} = \begin{cases} \frac{NTU}{1+NTU} & \text{for } C_{RAT} = 1 \\ \frac{1-e^{-NTU(1-C_{RAT})}}{1-C_{RAT}e^{-NTU(1-C_{RAT})}} & \text{for } C_{RAT} < 1 \end{cases} \tag{6.1}$$

where

ϵ_{CF} = Effectiveness of an ideal counterflow recuperator;
NTU = Number of Transfer Units, a dimensionless heat-exchanger size (–); and
C_{RAT} = Heat-capacity-rate ratio, the ratio of the heat-capacity rate of the compressed air to the heat-capacity rate of the exhaust (–).

(Equation 6.1 is derived later in this chapter.) Ideal conditions are:

1. Sufficiently fast core rotation;
2. No heat conduction through the core from the hot face to the cool face while the core is exposed to the flows;
3. No heat conduction while the core is under the seals; and
4. Uniform flow distribution over the core face.

The first three of these requirements for ideal conditions are the most relevant to the designer. To achieve a high effectiveness the designer would like a very fast core rotation rate; however, this would lead to excessive carry-over leakage of compressed air to the exhaust flow, which would reduce cycle thermal efficiency. The designer must be able to quantify the degree to which effectiveness will suffer due to only a modest core rotation rate, so the designer can make a proper trade-off of reduced heat transfer for less leakage. Similarly, highly compact regenerators are desirable for a small and lightweight engine; however, compact regenerators have thin cores with (usually[1]) a large amount of heat conduction from the from the hot face to the cool face, so the designer must be able to quantify the degree to which core conduction reduces regenerator effectiveness.

When one or more of the requirements for ideal conditions listed above are not met, then regenerator effectiveness is less than for an ideal counterflow recuperator. We quantify these reductions with "effects":

For slow core rotation;

$$\begin{aligned} \Delta_{ROT} &\equiv \frac{\epsilon_{ROT} - \epsilon_{CF}}{\epsilon_{CF}} \\ &\approx -\left(0.0018636NTU^2 + 0.39482NTU + 9.5909\right)^{-1} \times \\ &\quad \left(C_{ROT}^{-0.0025NTU+1.65}\right)^{-1}, \end{aligned} \tag{6.2}$$

where

Δ_{ROT} = Core-Rotation Effect (–); and
ϵ_{ROT} = Effectiveness of a regenerator with a finite core rotation rate (–); and
C_{ROT} = Dimensionless core rotation rate (–).

[1]An exception is regenerators with Mesh Screen Matrix (MSM) cores, for which heat conduction is negligible (see Chapters 4 and 5).

For axial conduction during flow exposures;

$$\Delta_{AC} \equiv \frac{\epsilon_{AC} - \epsilon_{CF}}{\epsilon_{CF}}$$
$$\approx -\left(\frac{1}{2 - C_{RAT}}\right) \times$$
$$\left(\frac{1}{1 + NTU - \kappa NTU^3/a^2} - \frac{1}{1 + NTU}\right) , \tag{6.3}$$

$$a \equiv 2\sqrt{NTU^2 + \frac{NTU}{\lambda}} ; \tag{6.4}$$

$$\kappa \equiv 4 - \left(\frac{8}{a}\right)\left(1 - \frac{1 - e^{-a}}{\sinh a}\right) ; \tag{6.5}$$

and

Δ_{AC} = Axial-Conduction Effect (–), which includes the effects of axial conduction but *not* finite core-rotation rate;
ϵ_{AC} = Effectiveness of a regenerator that rotates infinitely fast, but has axial conduction (–); and
λ = Core conduction parameter (–).

This axial-conduction effect was derived from a correlation proposed by Shah [61].

For axial conduction while the core is under the seals;

$$\Delta_S = \exp\{(-1.16C_{ROT} - 13.88) \times Fo^{(0.1245\lambda - 0.01798C_{ROT} + 0.9653)}\} - 1 , \tag{6.6}$$

where

$$Fo = \frac{SC\lambda}{2(1 - SC)C_{ROT}} \text{, a Fourier number.} \tag{6.7}$$

Using the three "effects" listed above, the effectiveness of gas-turbine regenerators can be calculated to a high degree of accuracy using the following correlation:

$$\epsilon_{ROT,AC,S} \approx (1 + \Delta_{ROT})(1 + \Delta_{AC})(1 + \Delta_S)\epsilon_{CF} - \Delta_S\epsilon_{ROT,AC,S}(SC \to 1) , \tag{6.8}$$

where

$$\epsilon_{ROT,AC,S}(SC \to 1) \approx 0.50505 + \frac{0.257}{C_{ROT} - 0.5} . \tag{6.9}$$

This section has summarized how to calculate regenerator effectiveness using correlations of data calculated using sophisticated numerical techniques. The rest of this section on heat transfer describes in detail how to employ these numerical techniques. First section 6.1.2 describes the physical heat-transfer phenomena in regenerators. Chapter 1 showed that low porosity and flow non-uniformity can reduce the effectiveness (ϵ) of regenerators. Section 6.1.10 quantifies reductions in effectiveness due to low porosity. Sections 6.1.12 and 6.1.13 quantify effectiveness reductions due to two causes of flow non-uniformity: (1) arbitrary seal shape; and (2) discontinuous rotation. Chapter 1 also showed that effectiveness is reduced during transient operation. Section 6.1.11 quantifies the effects of transient operation.

6.1.2 Physical Phenomena

Figure 1.6 shows a schematic of a passage tube of a regenerator core. The passage tube initially has a uniform temperature distribution, T_{R0}. Then, fluid enters the passage with a different temperature, T_i, and heats or cools the passage-tube walls through forced convection. The fluid flow's velocity profile develops after a hydraulic development length, x_{Hcrit}. The fluid flow's dimensionless temperature profile,

$$\Theta \equiv \frac{T_0 - T_1}{T_0 - T_m} \tag{6.10}$$

(where T_0 is the temperature at the fluid-solid interface; T_1 is the temperature in the fluid domain; and T_m is the mixed-mean temperature in the fluid domain), develops after a thermal development length, x_{THcrit}. Heat conducts in all three directions in the passage tube. In addition to conduction associated with convection, heat conducts axially in the fluid. Heat also radiates along the length of the passage.

Between exposures to the two fluid flows, passage tubes rotate under seals. While a passage tube is under a seal, a small amount of fluid flows into (or out of) the passage due to direct seal leakage. When the passage tube is under seals, heat conducts in all three directions. During seal coverage there is free convection. Also, heat radiates along the length of the passage during seal coverage.

6.1.3 Simplifying Assumptions

We simplify the analysis of regenerators with the following assumptions:

1. Negligible conduction through passage-tube boundaries;
2. Laminar flows;
3. Negligible axial conduction in the fluid;
4. Hydraulically and thermally fully developed flows;
5. Constant properties of each fluid and of the core material;
6. Negligible radiation between surfaces of the regenerator core;

7. Negligible heat transfer from passage-tube walls at the ends of passages;
8. Negligible free convection under seals;
9. Negligible flows into and out of passages under seals; and
10. The heat capacity of the core material is much greater than the heat capacities of either of the fluids in the passages.

The first assumption is valid in a regenerator if the passages are sufficiently small and the flows are sufficiently uniform over the flow areas. The second assumption is valid for low Reynolds numbers: $Re_{D_H} < 2300$ (see Potter and Foss [57]). For gas-turbine applications, typically $50 < Re_{D_H} < 200$. The third assumption is valid for large Péclét numbers: $Pe_{D_H} > 10$ (see Shah and London [63]). For air flows, the Prandlt number is $Pr \approx 0.7$. Therefore, for air-breathing regenerative gas turbines, the Péclét number typically lies in the range $35 < Pe_{D_H} < 140$.

The fourth assumption is accurate if the development lengths are a sufficiently small portion of the total passage length. Kays and Crawford [36] give the following equation for the hydraulic development length, x_{Hcrit}:

$$\frac{x_{Hcrit}}{D_H} = \frac{Re_{D_H}}{20} , \tag{6.11}$$

where D_H is the hydraulic diameter. A typical hydraulic diameter for a passage in a gas-turbine regenerator is $D_H = 0.5$ mm. A typical Reynolds number is $Re_{D_H} = 50$. A typical core length is in the range L =50–150 mm. Therefore,

$$\frac{x_{Hcrit}}{L} \approx 1\text{–}3\% . \tag{6.12}$$

Kays and Crawford [36] give the following equation for the thermal development length:

$$\frac{x_{THcrit}}{D_H} \approx (0.05) Pe_{D_H} . \tag{6.13}$$

Therefore,

$$0.6\% \leq \frac{x_{THcrit}}{L} \leq 2\% . \tag{6.14}$$

In a typical gas-turbine regenerator, the development lengths are sufficiently small to assume hydraulically and thermally fully developed flows.

Kays and London [37] discuss the effects of temperature-dependent fluid properties. The fifth assumption is accurate if the properties are calculated at average temperatures. The sixth assumption typically is valid for gas-turbine applications because the views from the hot passage portions to the cool passage portions typically are poor.

The seventh assumption is valid for gas-turbine applications because the heat-transfer areas occupied by passage-tube ends are typically small portions of the total heat-transfer areas, especially for high-porosity cores. The eighth assumption is valid for the typically small passage diameters associated with gas-turbine applications. The ninth assumption is valid for low direct seal leakage. The tenth assumption is valid for gas flows through ceramic or metallic cores.

6.1.4 Governing Equations

This section derives governing equations from the simplified analysis given in the last section. This section derives the most general equations that are considered for calculating the heat-transfer performance of gas-turbine regenerators. Consider the isolated passage tube shown in Figure 6.1. Conservation of energy for the indicated element of length Δx when it is exposed to the C_N-side flow (usually the compressed-air flow in gas-turbine regenerators) yields

$$\begin{aligned}&\frac{C_N}{N_{pass_N}}\left[T_N-\left(T_N+\frac{\partial T_N}{\partial x}\Delta x\right)\right]\\&-\frac{k_R As_N}{N_{pass_N}}\left[\frac{\partial T_R}{\partial x}-\left(\frac{\partial T_R}{\partial x}+\frac{\partial^2 T_R}{\partial x^2}\Delta x\right)\right]\\&=\frac{(mc)_R}{N_{pass_N}}\left(\frac{\Delta x}{L}\right)\left(\frac{As_N}{As_\Sigma}\right)\frac{\partial T_R}{\partial t}\ ,\end{aligned} \tag{6.15}$$

where N_{pass_N} is the number of core passages on the C_N side. Equation 6.15 can be rewritten as

$$C_{RAT}\frac{\partial T_N}{\partial \zeta}+\frac{(1-SC)As'}{(1+As')}\frac{\partial T_R}{\partial t_X^\star}-\frac{\lambda As' C_{RAT}}{(1+As')}\frac{\partial^2 T_R}{\partial \zeta^2}=0\ . \tag{6.16}$$

Newton's law of convection gives the amount of heat gained by the C_N-side flow through the passage element:

$$\frac{C_N}{N_{pass_N}}\left[\left(T_N+\frac{\partial T_N}{\partial x}\Delta x\right)-T_N\right]=\frac{(hA)_N}{N_{pass_N}}\left(\frac{\Delta x}{L}\right)[T_R-T_N]\ . \tag{6.17}$$

This heat gain of the C_N-side flow produces the following core-material temperature response.

$$-\frac{(mc)_R}{N_{pass_N}}\left(\frac{\Delta x}{L}\right)\left(\frac{As_N}{As_\Sigma}\right)\frac{\partial T_R}{\partial t}=\frac{(hA)_N}{N_{pass_N}}\left(\frac{\Delta x}{L}\right)[T_R-T_N]\ . \tag{6.18}$$

Equations 6.17 and 6.18 can be rewritten as

$$T_R=T_N+\frac{1}{NTU[1+(hA)']}\frac{\partial T_N}{\partial \zeta}\ , \tag{6.19}$$

and

$$T_N=T_R+\frac{As'(1-SC)}{(1+As')NTU[1+(hA)']C_{RAT}}\frac{\partial T_R}{\partial t_X^\star}\ . \tag{6.20}$$

Equation 6.19 can be substituted into Equation 6.16 to yield

$$\begin{aligned}&C_{RAT}\frac{\partial T_N}{\partial \zeta}+\frac{(1-SC)As'}{(1+As')}\frac{\partial T_N}{\partial t_X^\star}\\&+\frac{(1-SC)As'}{(1+As')NTU[1+(hA)']}\frac{\partial}{\partial \zeta}\left(\frac{\partial T_N}{\partial t_X^\star}\right)\end{aligned}$$

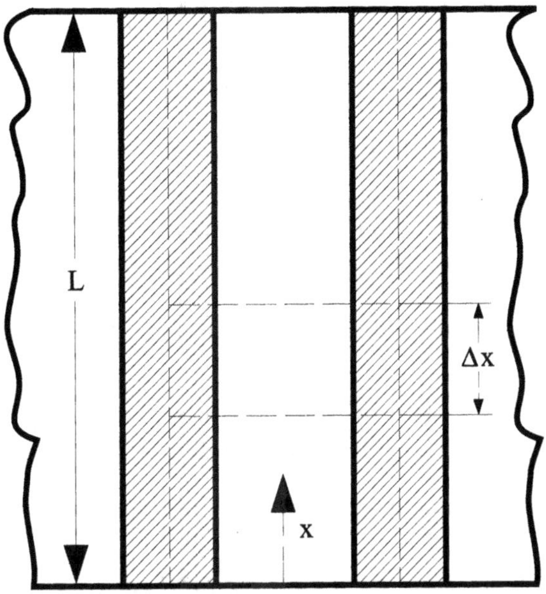

Figure 6.1: Isolated Regenerator Passage Tube.

$$
\begin{aligned}
&-\frac{\lambda As' C_{RAT}}{(1+As')}\frac{\partial^2 T_N}{\partial \zeta^2} \\
&-\frac{\lambda As' C_{RAT}}{(1+As')NTU[1+(hA)']}\frac{\partial^3 T_N}{\partial \zeta^3} = 0 \,. \qquad (6.21)
\end{aligned}
$$

Equation 6.20 can be substituted into Equation 6.16 to yield

$$
\begin{aligned}
&C_{RAT}\frac{\partial T_R}{\partial \zeta} + \frac{(1-SC)As'}{(1+As')}\frac{\partial T_R}{\partial t^{\star}_X} \\
&+\frac{(1-SC)As'}{(1+As')NTU[1+(hA)']}\frac{\partial}{\partial \zeta}\left(\frac{\partial T_R}{\partial t^{\star}_X}\right) \\
&-\frac{\lambda As' C_{RAT}}{(1+As')}\frac{\partial^2 T_R}{\partial \zeta^2} = 0 \,. \qquad (6.22)
\end{aligned}
$$

When the passage tube is under a seal,

$$
\begin{aligned}
&-\frac{k_R As_S}{N_{pass_S}}\left[\frac{\partial T_R}{\partial x} - \left(\frac{\partial T_R}{\partial x} + \frac{\partial^2 T_R}{\partial x^2}\Delta x\right)\right] \\
&= \frac{(mc)_R}{N_{pass_S}}\left(\frac{\Delta x}{L}\right)\left(\frac{As_S}{As_\Sigma}\right)\frac{\partial T_R}{\partial t} \,, \qquad (6.23)
\end{aligned}
$$

where N_{pass_S} is the number of passages under both seals. Equation 6.23 can be rewritten as

$$
\frac{\partial^2 T_R}{\partial \zeta^2} = \frac{(1-SC)}{C_{RAT}\lambda}\frac{\partial T_R}{\partial t^{\star}_X} \,. \qquad (6.24)
$$

When the passage tube is exposed to the C_X-side flow,

$$\begin{aligned}
&\frac{C_X}{N_{pass_X}}\left[\left(T_X+\frac{\partial T_X}{\partial x}\Delta x\right)-T_X\right] \\
&-\frac{k_R As_X}{N_{pass_X}}\left[\frac{\partial T_R}{\partial x}-\left(\frac{\partial T_R}{\partial x}+\frac{\partial^2 T_R}{\partial x^2}\Delta x\right)\right] \\
&=\frac{(mc)_R}{N_{pass_X}}\left(\frac{\Delta x}{L}\right)\left(\frac{As_X}{As_\Sigma}\right)\frac{\partial T_R}{\partial t}\ ,
\end{aligned} \tag{6.25}$$

where N_{pass_X} is the number of passages on the C_X side. Equation 6.25 can be rewritten as

$$-\frac{\partial T_X}{\partial \zeta}+\frac{(1-SC)}{(1+As')}\frac{\partial T_R}{\partial t_X^\star}-\frac{C_{RAT}\lambda}{(1+As')}\frac{\partial^2 T_R}{\partial \zeta^2}=0\ . \tag{6.26}$$

Newton's law of convection gives the amount of heat lost by the C_X-side flow through the passage element:

$$\frac{C_X}{N_{pass_X}}\left[\left(T_X+\frac{\partial T_X}{\partial x}\Delta x\right)-T_X\right]=\frac{(hA)_X}{N_{pass_X}}\left(\frac{\Delta x}{L}\right)[T_X-T_R]\ . \tag{6.27}$$

This heat lost by the C_X-side flow produces the following core-material temperature response:

$$\frac{(mc)_R}{N_{pass_X}}\left(\frac{\Delta x}{L}\right)\left(\frac{As_X}{As_\Sigma}\right)\frac{\partial T_R}{\partial t}=\frac{(hA)_X}{N_{pass_X}}\left(\frac{\Delta x}{L}\right)[T_X-T_R]\ . \tag{6.28}$$

Equations 6.27 and 6.28 can be rewritten as

$$T_R=T_X-\frac{(hA)'}{NTU[1+(hA)']C_{RAT}}\frac{\partial T_X}{\partial \zeta}\ , \tag{6.29}$$

and

$$T_X=T_R+\frac{(1-SC)(hA)'}{(1+As')NTU[1+(hA)']C_{RAT}}\frac{\partial T_R}{\partial t_X^\star}\ . \tag{6.30}$$

Equation 6.29 can be substituted into Equation 6.26 to yield

$$\begin{aligned}
&-\frac{\partial T_X}{\partial \zeta}+\frac{(1-SC)}{(1+As')}\frac{\partial T_X}{\partial t_X^\star} \\
&-\frac{(1-SC)(hA)'}{(1+As')NTU[1+(hA)']C_{RAT}}\frac{\partial}{\partial \zeta}\left(\frac{\partial T_X}{\partial t_X^\star}\right) \\
&-\frac{C_{RAT}\lambda}{(1+As')}\frac{\partial^2 T_X}{\partial \zeta^2} \\
&+\frac{\lambda(hA)'}{(1+As')NTU[1+(hA)']}\frac{\partial^3 T_X}{\partial \zeta^3}\ =\ 0\ .
\end{aligned} \tag{6.31}$$

Equation 6.30 can be substituted into Equation 6.26 to yield

$$-\frac{\partial T_R}{\partial \zeta} + \frac{(1-SC)}{(1+As')}\frac{\partial T_R}{\partial t_X^\star} - \frac{(1-SC)(hA)'}{(1+As')NTU[1+(hA)']C_{RAT}}\frac{\partial}{\partial \zeta}\left(\frac{\partial T_R}{\partial t_X^\star}\right) - \frac{C_{RAT}\lambda}{(1+As')}\frac{\partial^2 T_R}{\partial \zeta^2} = 0\,. \tag{6.32}$$

Some boundary conditions associated with the above equations are

1. $\partial T_R/\partial \zeta|_{\zeta=0,1} = 0$;
2. $\partial T_{N,X}/\partial \zeta|_{\zeta=0,1} = 0$;
3. $T_R(t_X^\star = 0; \zeta) = T_{R0}$;
4. $T_N(t_X^\star; \zeta = 0) = T_{Ni}$; and
5. $T_X(t_X^\star; \zeta = 1) = T_{Xi}$.

The first boundary conditions result from the assumption of no heat transfer from passage tube ends. The second boundary conditions result from the assumption of no axial conduction in the fluid. In addition to the numbered boundary conditions given above,

1. The initial core temperature profile when a passage enters the C_N side is equal to the final temperature profile when the passage exits from under the preceding seal;
2. The final core temperature profile when a passage exits from the C_N side is equal to the initial temperature profile when the passage enters under the proceding seal;
3. The initial core temperature profile when a passage enters the C_X side is equal to the final temperature profile when the passage exits from under the preceding seal; and
4. The final core temperature profile when a passage exits from the C_X side is equal to the initial temperature profile when the passage enters under the proceding seal.

A passage is exposed to the C_N-side flow for

$$\tau_{X_N}^\star = \frac{As'(1-SC)}{(1+As')C_{ROT}C_{RAT}}\,. \tag{6.33}$$

A passage is under one of the two seals for

$$\frac{\tau_{X_S}^\star}{2} = \frac{SC}{2C_{ROT}C_{RAT}}\,. \tag{6.34}$$

A passage is exposed to the C_X-side flow for

$$\tau^\star_{X_X} = \frac{(1 - SC)}{(1 + As')C_{ROT}C_{RAT}} \;. \qquad (6.35)$$

Equations 6.21, 6.22, 6.24, 6.31, and 6.32 subject to the given boundary conditions are appropriate for calculation of transient temperature and effectiveness responses. For steady-state calculations, we use the following substitutions[2]. For the C_N side,

$$\xi_N \equiv \frac{t^\star_X(1 + As')C_{RAT}C_{ROT}}{As'(1 - SC)} \;. \qquad (6.36)$$

For passage tubes under the seals,

$$\xi_{S1,2} \equiv \frac{2t^\star_X C_{RAT}C_{ROT}}{SC} \;. \qquad (6.37)$$

For the C_X side,

$$\xi_X \equiv \frac{t^\star_X(1 + As')C_{RAT}C_{ROT}}{(1 - SC)} \;. \qquad (6.38)$$

Then Equations 6.21, 6.22, 6.24, 6.31, and 6.32 become

$$\frac{\partial T_N}{\partial \zeta} + C_{ROT}\frac{\partial T_N}{\partial \xi_N} + \frac{C_{ROT}}{NTU[1 + (hA)']}\frac{\partial}{\partial \zeta}\left(\frac{\partial T_N}{\partial \xi_N}\right) - \frac{\lambda As'}{(1 + As')}\frac{\partial^2 T_N}{\partial \zeta^2} - \frac{\lambda As'}{(1 + As')NTU[1 + (hA)']}\frac{\partial^3 T_N}{\partial \zeta^3} = 0 \;; \qquad (6.39)$$

$$\frac{\partial T_R}{\partial \zeta} + C_{ROT}\frac{\partial T_R}{\partial \xi_N} + \frac{C_{ROT}}{NTU[1 + (hA)']}\frac{\partial}{\partial \zeta}\left(\frac{\partial T_R}{\partial \xi_N}\right) - \frac{\lambda As'}{(1 + As')}\frac{\partial^2 T_R}{\partial \zeta^2} = 0 \;; \qquad (6.40)$$

$$\frac{\partial^2 T_R}{\partial \zeta^2} = \frac{2(1 - SC)C_{ROT}}{\lambda SC}\frac{\partial T_R}{\partial \xi_{S1,2}} \;; \qquad (6.41)$$

[2]For steady-state calculations, only differences in time are significant. For simplicity in the definitions, we set the time to zero when the core material enters the side or seal considered. Thus, $0 \le \xi \le 1$.

$$
\begin{aligned}
-\frac{\partial T_X}{\partial \zeta} + C_{RAT}C_{ROT}\frac{\partial T_X}{\partial \xi_X} \\
-\frac{(hA)'C_{ROT}}{NTU[1+(hA)']}\frac{\partial}{\partial \zeta}\left(\frac{\partial T_X}{\partial \xi_X}\right) \\
-\frac{C_{RAT}\lambda}{(1+As')}\frac{\partial^2 T_X}{\partial \zeta^2} \\
+\frac{\lambda(hA)'}{(1+As')NTU[1+(hA)']}\frac{\partial^3 T_X}{\partial \zeta^3} &= 0\,; \qquad (6.42)
\end{aligned}
$$

and

$$
\begin{aligned}
-\frac{\partial T_R}{\partial \zeta} + C_{RAT}C_{ROT}\frac{\partial T_R}{\partial \xi_X} \\
-\frac{(hA)'C_{ROT}}{NTU[1+(hA)']}\frac{\partial}{\partial \zeta}\left(\frac{\partial T_R}{\partial \xi_X}\right) \\
-\frac{C_{RAT}\lambda}{(1+As')}\frac{\partial^2 T_R}{\partial \zeta^2} &= 0\,. \qquad (6.43)
\end{aligned}
$$

The boundary conditions associated with these equations are

1. $\partial T_R/\partial\zeta|_{\zeta=0,1} = 0$;
2. $\partial T_{N,X}/\partial\zeta|_{\zeta=0,1} = 0$;
3. $T_N(\xi_N; \zeta = 0) = T_{Ni}$;
4. $T_X(\xi_X; \zeta = 1) = T_{Xi}$;
5. $T_R(\xi_N = 0; \zeta) = T_R(\xi_{S2} = 1; \zeta)$;
6. $T_R(\xi_{S1} = 0; \zeta) = T_R(\xi_N = 1; \zeta)$;
7. $T_R(\xi_X = 0; \zeta) = T_R(\xi_{S1} = 1; \zeta)$; and
8. $T_R(\xi_{S2} = 0; \zeta) = T_R(\xi_X = 1; \zeta)$.

The equations given above are applicable for high porosity cores. Regenerators with parallel-plate passages could have low porosities (see Section 1.2.7). Therefore, we give governing equations that we use to determine the effects of low porosity on the effectiveness of parallel-plate regenerators.

Figure 1.8 shows a schematic of a passage tube in a parallel-plate regenerator. Conservation of momentum produces

$$\frac{d^2u}{dy_1^2} = \frac{1}{\mu}\frac{dP}{dx}\,, \qquad (6.44)$$

where u is the flow velocity in the x direction; μ is the fluid's viscosity; and P is the fluid's pressure. We solve Equation 6.44 in closed form to obtain the following velocity profile for laminar flow between parallel plates:

$$u^\star = \left[\frac{3}{2}\right]\left[1 - y_1^{\star^2}\right]\,, \qquad (6.45)$$

where

$$u^\star \equiv \frac{u}{U} \, , \tag{6.46}$$

and

$$y_1^\star \equiv \frac{y_1}{Y} \, . \tag{6.47}$$

Under the relevant assumptions given in Section 6.1.3, the appropriate energy equation for the fluid flow is

$$\frac{\partial^2 T_1}{\partial y_1^2} = \left(\frac{u}{\alpha}\right) \frac{\partial T_1}{\partial x} \, . \tag{6.48}$$

It is assumed that the flow has a fully developed temperature profile. This means that the nondimensional temperature,

$$\Theta \equiv \frac{T_0 - T_1}{T_0 - T_m} \, , \tag{6.49}$$

has a profile that is invariant with x, or $\partial\Theta/\partial x = 0$. In a typical gas-turbine regenerator, a nearly linear axial temperature profile is established. This means that the boundary condition at the fluid/solid interface is a constant heat flux with axial position. This boundary condition produces a fully-developed dimensionless-temperature profile after a certain axial thermal-development distance. Section 6.1.3 showed that this distance typically is a small fraction of the total core thickness.

A consequence of a fully-developed temperature profile and a constant heat-flux boundary condition is

$$\frac{\partial T_1}{\partial x} = \frac{dT_m}{dx} \, . \tag{6.50}$$

Therefore, Equation 6.48 becomes

$$\frac{\partial^2 T_1}{\partial y_1^2} = \left(\frac{u}{\alpha}\right) \frac{dT_m}{dx} \, , \tag{6.51}$$

or

$$\frac{\partial^2 \Theta}{\partial y_1^{\star 2}} = \left(\frac{-u^\star}{4}\right) Nu_H \, . \tag{6.52}$$

Under the assumption of constant properties (which is discussed in Section 6.1.3), the appropriate energy equation in the core-material domain is

$$\frac{\partial^2 T_2}{\partial y_2^2} = \left(\frac{1}{\alpha_R}\right) \frac{\partial T_2}{\partial t} \, , \tag{6.53}$$

which is a one-dimensional heat-diffusion equation. Equation 6.53 can be expressed as

$$\frac{\partial^2 T_2}{\partial y_2^{\star 2}} = \frac{\partial T_2}{\partial Fo_H} \, , \tag{6.54}$$

where

$$y_2^\star \equiv \frac{y_2}{H} , \qquad (6.55)$$

and

$$Fo_H \equiv \frac{\alpha_R t}{H^2} . \qquad (6.56)$$

This section has derived the governing equations that are used to quantify the heat-transfer performance of gas-turbine regenerators. The following sections give solutions to these governing equations under decreasingly stringent assumptions. Thus, the following sections quantify heat-transfer performance in gas-turbine regenerators.

6.1.5 Infinite Core-Rotation Rate

If the core of regenerator rotates sufficiently fast, it may be possible to calculate the effectiveness (ϵ) of the regenerator using Equation 6.1. This equation is valid if

1. The regenerator core rotates infinitely fast ($C_{ROT} > 5$ usually leads to negligible errors);
2. The regenerator core has zero conduction in the direction of fluid flows ($\lambda = 0$); and
3. The assumptions given in Section 6.1.3 hold.

Figure 6.2 gives plots of Equation 6.1. Figure 6.2 shows that increasing the size of the regenerator core (increasing the Number of Transfer Units, NTU) has diminishing returns in improving effectiveness (ϵ). In addition, effectiveness is greatest for highly unequal heat-capacity rates.

The regenerator designer is most interested in determining the dimensionless core size (NTU) that is required to achieve the specified effectiveness (ϵ). To facilitate this task, we invert Equation 6.1 to obtain the following expressions for NTU. For equal heat-capacity rates ($C_{RAT} = 1$),

$$NTU = \frac{\epsilon}{1-\epsilon} ; \qquad (6.57)$$

and for unequal heat-capacity rates ($C_{RAT} < 1$),

$$NTU = \frac{\ln\left[\frac{1-C_{RAT}\epsilon}{1-\epsilon}\right]}{1-C_{RAT}} . \qquad (6.58)$$

Of course these equations are valid only if the three criteria listed above are met.

The regenerator designer is sometimes interested in the temperature distribution through the regenerator core, such as when the designer considers the

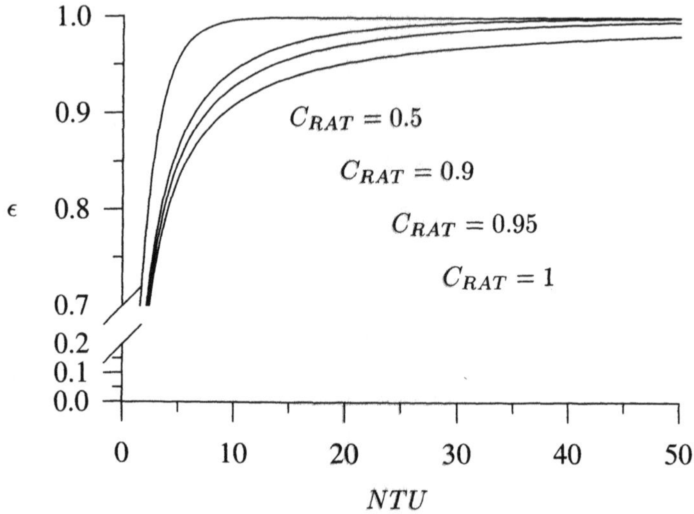

Figure 6.2: Effect of Dimensionless Core Size (NTU) on Effectiveness (ϵ).

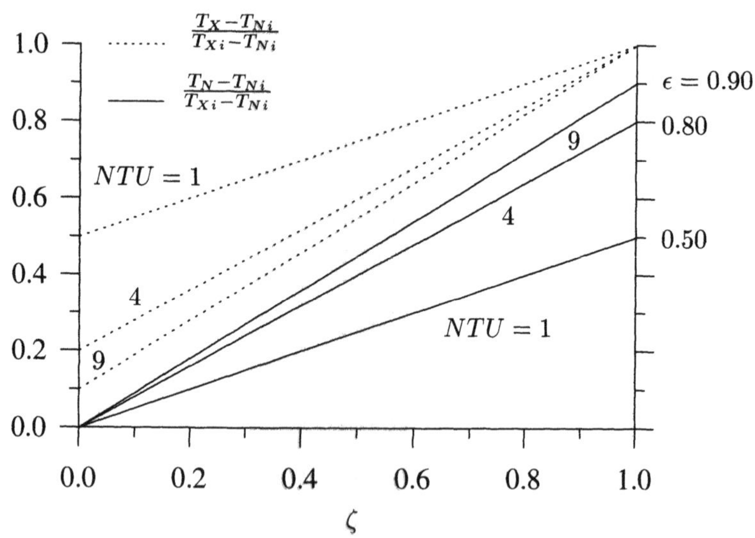

Figure 6.3: Temperature Distributions through a Regenerator Core for Flows with Equal Heat-Capacity Rates ($C_{RAT} = 1$).

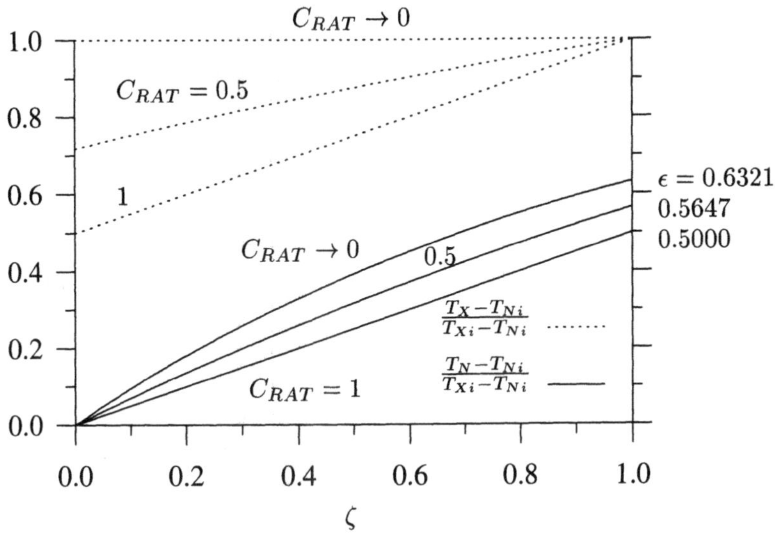

Figure 6.4: Temperature Distributions through a Regenerator Core ($NTU = 1$).

material integrity of the core. Figures 6.3 and 6.4 show the temperature distributions through regenerator cores for infinite core-rotation rate ($C_{ROT} = \infty$) and the other assumptions of this section.

The remainder of this section describes how Equation 6.1 is derived from the governing equations given in Section 6.1.4. Also, expressions for the temperature distribution through the regenerator core are derived.

From Equation 6.16 with zero conduction ($\lambda = 0$) and the ξ_N substitution,

$$\frac{\partial T_N}{\partial \zeta} = -C_{ROT} \frac{\partial T_R}{\partial \xi_N} \ . \tag{6.59}$$

Similarly, for the C_X side,

$$\frac{\partial T_X}{\partial \zeta} = C_{RAT} C_{ROT} \frac{\partial T_R}{\partial \xi_X} \ . \tag{6.60}$$

These equations indicate that for fast core-rotation rates (large C_{ROT}), temperature gradients in the core-rotation (ξ_N and ξ_X) directions are small compared with temperature gradients in the flow (ζ) direction. Therefore, the following approximations can be used for sufficiently fast core-rotation rates (large C_{ROT}):

$$\frac{\partial T_R}{\partial \xi_N} \approx T_R(\zeta; \xi_N = 1) - T_R(\zeta; \xi_N = 0) \ ; \tag{6.61}$$

$$\frac{\partial T_R}{\partial \xi_X} \approx T_R(\zeta; \xi_X = 1) - T_R(\zeta; \xi_X = 0) \ ; \tag{6.62}$$

$$T_R(\zeta; \xi_N) \approx \frac{T_R(\zeta; \xi_N = 1) + T_R(\zeta; \xi_N = 0)}{2} \ ; \text{ and} \tag{6.63}$$

$$T_R(\zeta;\xi_X) \approx \frac{T_R(\zeta;\xi_X=1)+T_R(\zeta;\xi_X=0)}{2} . \tag{6.64}$$

We substitute these equations into Equations 6.59, 6.60, 6.20, and 6.30 (with the ξ_N and ξ_X substitutions) to yield

$$\frac{\partial T_N}{\partial \zeta} = -C_{ROT}\left[T_R(\zeta;\xi_N=1)-T_R(\zeta;\xi_N=0)\right] ; \tag{6.65}$$

$$\frac{\partial T_X}{\partial \zeta} = -C_{RAT}C_{ROT}\left[T_R(\zeta;\xi_X=1)-T_R(\zeta;\xi_X=0)\right] ; \tag{6.66}$$

$$\begin{aligned} T_N \;=\; & \frac{T_R(\zeta;\xi_N=1)+T_R(\zeta;\xi_N=0)}{2} \\ & +\frac{C_{ROT}}{NTU[1+(hA)']} \\ & [T_R(\zeta;\xi_N=1)-T_R(\zeta;\xi_N=0)] \text{ ; and} \end{aligned} \tag{6.67}$$

$$\begin{aligned} T_X \;=\; & \frac{T_R(\zeta;\xi_X=1)+T_R(\zeta;\xi_X=0)}{2} \\ & +\frac{(hA)'C_{ROT}}{NTU[1+(hA)']}\left[T_R(\zeta;\xi_X=1)-T_R(\zeta;\xi_X=0)\right] . \end{aligned} \tag{6.68}$$

The assumption of zero axial conduction precludes heat diffusion under seals. Therefore, $T_R(\zeta;\xi_N=0)=T_R(\zeta;\xi_X=1)$ and $T_R(\zeta;\xi_N=1)=T_R(\zeta;\xi_X=0)$. Thus,

$$\frac{\partial}{\partial \zeta}[T_X-T_N] = -NTU[1-C_{RAT}][T_X-T_N] . \tag{6.69}$$

The boundary conditions associated with this equation are

1. $[T_X-T_N](\zeta=0)=T_{Xo}-T_{Ni}$ and
2. $[T_X-T_N](\zeta=1)=T_{Xi}-T_{No}$.

The solution to Equation 6.69 subject to the above boundary conditions is

$$\left[\frac{T_X-T_N}{T_{Xi}-T_{Ni}}\right](\zeta) = \exp\left\{-NTU[1-C_{RAT}]\zeta\right\}[1-\epsilon C_{RAT}] , \tag{6.70}$$

where we calculate the effectiveness (ϵ) using Equation 6.1. Also,

$$\frac{T_N-T_{Ni}}{T_{Xi}-T_{Ni}} = \frac{1-\exp\left\{-NTU[1-C_{RAT}]\zeta\right\}}{1-C_{RAT}\exp\left\{-NTU[1-C_{RAT}]\right\}} , \tag{6.71}$$

and

$$\frac{T_X-T_{Ni}}{T_{Xi}-T_{Ni}} = \frac{1-C_{RAT}\exp\left\{-NTU[1-C_{RAT}]\zeta\right\}}{1-C_{RAT}\exp\left\{-NTU[1-C_{RAT}]\right\}} . \tag{6.72}$$

Equation 6.69 and the associated boundary conditions are identical to those of a counterflow recuperator. Therefore, the solutions are identical. See Kays and London [38] for a derivation of the equations for a counterflow recuperator.

The solution is simplified as the heat-capacity-rate ratio (C_{RAT}) approaches its limits. As $C_{RAT} \to 0$,

$$\epsilon \to 1 - \exp\{-NTU\} \ ; \tag{6.73}$$

$$\frac{T_N - T_{Ni}}{T_{Xi} - T_{Ni}} \to 1 - \exp\{-NTU\zeta\} \ ; \text{ and} \tag{6.74}$$

$$\frac{T_X - T_{Ni}}{T_{Xi} - T_{Ni}} \to 1 \ . \tag{6.75}$$

For $C_{RAT} = 1$, we calculate effectiveness (ϵ) using Equation 6.1;

$$\frac{T_N - T_{Ni}}{T_{Xi} - T_{Ni}} = \frac{NTU\zeta}{1 + NTU} \ ; \tag{6.76}$$

and

$$\frac{T_X - T_{Ni}}{T_{Xi} - T_{Ni}} = \frac{1 + NTU\zeta}{1 + NTU} \ . \tag{6.77}$$

This section has shown how to calculate regenerator effectiveness (ϵ) from the dimensionless core size (NTU) and vice versa for infinite core-rotation rate ($C_{ROT} = \infty$) and zero core conduction ($\lambda = 0$). Also, this section has shown how to calculate the temperature distribution through the core. The next section relaxes the assumption of infinite core-rotation rate.

6.1.6 Finite Core-Rotation Rate

Analytic expressions for effectiveness are obtainable only for very fast core-rotation rates ($C_{ROT} \to \infty$). Fast core-rotation rates lead to high carry-over leakage (see Section 6.3). Therefore, regenerator designers usually choose core-rotation rates that are slow enough to invalidate the analytic effectiveness expressions given in the last section. For slow core-rotation rates ($C_{ROT} < 5$), effectiveness is reduced.

Lambertson [40] quantified the effects of slow core-rotation on regenerator effectiveness. Figures 6.5–6.7 shows some of his results and some of the results obtained by Bahnke and Howard [2] using similar methods. The plots show that dimensionless core-rotation rates greater than about five ($C_{ROT} > 5$) produce effectivenesses that are almost equal to those for infinite core-rotation rates ($C_{ROT} = \infty$). For slower core-rotation rates, however, effectiveness is significantly lower.

Lambertson [40] solved the equations given in Section 6.1.4 under the assumptions of zero axial conduction ($\lambda = 0$) and steady-state operation. He used a finite-difference numerical-integration algorithm. He gives calculated effectiveness data for the following ranges of heat-transfer parameters.

1. $1 \leq NTU \leq 10$;

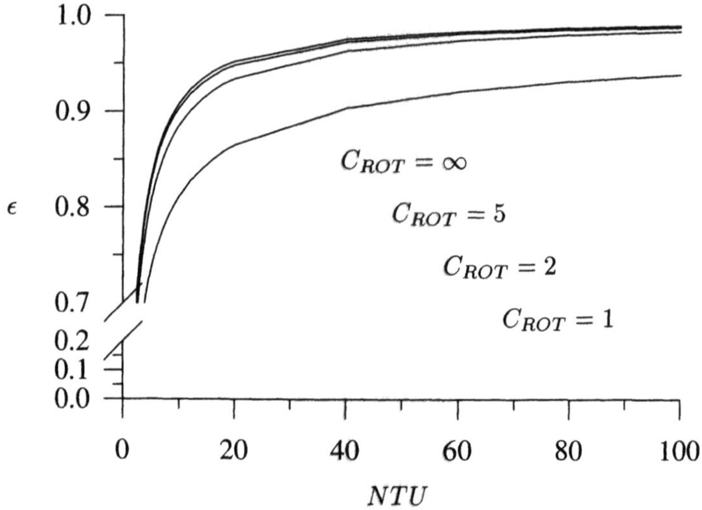

Figure 6.5: Effectiveness for Finite Core-Rotation Rates ($C_{RAT} = 1$ and $(hA)' = 1$).

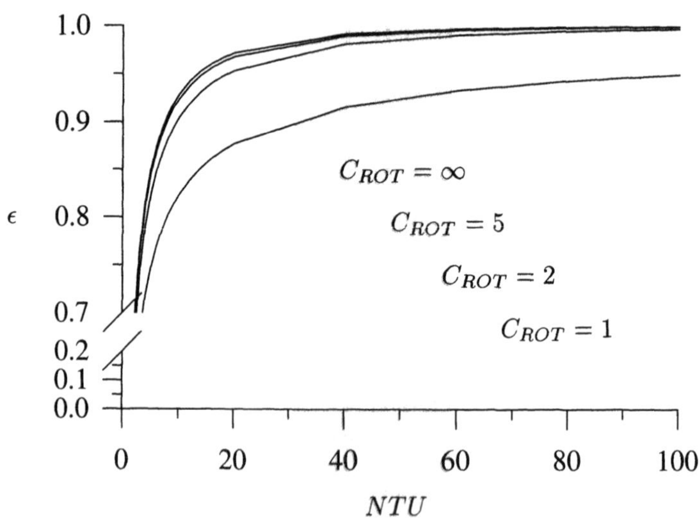

Figure 6.6: Effectiveness for Finite Core-Rotation Rates ($C_{RAT} = 0.9$ and $(hA)' = 1$).

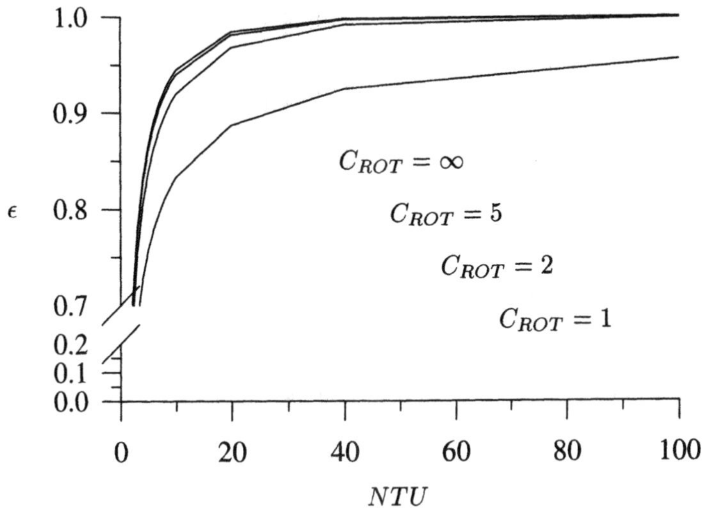

Figure 6.7: Effectiveness for Finite Core-Rotation Rates ($C_{RAT} = 0.95$ and $(hA)' = 1$).

2. $1 \leq C_{ROT} \leq \infty$;
3. $0.1 \leq C_{RAT} \leq 1$;
4. $0.25 \leq (hA)' \leq 1$; and
5. $\lambda = 0$.

He also gives the following rotation effect[3].

$$\Delta_{ROT} = -\frac{1}{9C_{ROT}^{1.87}} . \tag{6.78}$$

With this rotation effect, effectiveness can be calculated to within 1% for the following ranges of heat-transfer parameters:

1. $1 \leq NTU \leq 9$;
2. $1.25 \leq C_{ROT} \leq 5$;
3. $0.9 \leq C_{RAT} \leq 1$;
4. $0.25 \leq (hA)' \leq 1$; and
5. $\lambda = 0$.

Kays and London [37] modify the exponent in the rotation effect from 1.87 to 1.93, and they recommend the use of the resulting rotation effect for $\epsilon \leq 90\%$. The authors recommend the rotation effect given by Equation 6.2.

[3]Lambertson [40] defined his rotation effect with an opposite sign. In this book, "effects" are defined to be negative for effectiveness reductions.

6.1.7 Axial Conduction during Flow Exposures

Small passage dimensions make regenerator cores highly compact because they reduce the thickness of the core that is required to achieve a specified effectiveness (see Chapter 5). A thin core, however, has at least two disadvantages:

1. Thin cores have high conduction in the direction of the fluid flows, which reduces effectiveness; and

2. Thin cores have high thermal stresses.

These factors place lower bounds on core thickness.

Bahnke and Howard [2] quantified the effects of core conduction on effectiveness; however, their calculations neglected the effects of seal width, which the next section considers. The principal result of their analyses is that effectiveness is reduced for high values of the dimensionless core-conduction parameter, λ. Banhke and Howard [2] used a finite-difference numerical-integration algorithm. They give calculated effectiveness data for

1. $1 \leq NTU \leq 100$;

2. $1 \leq C_{ROT} \leq \infty$;

3. $0.9 \leq C_{RAT} \leq 1$;

4. $0.25 \leq (hA)' \leq 1$;

5. $0 \leq \lambda \leq 0.32$;

6. $0.25 \leq As' \leq 1$; and

7. $SC = 0$.

The rest of this section describes correlations of the effects of core conduction.

Kays and London [37] give the following approximation for the effects of core conduction:

$$\Delta_{AC} \equiv \frac{\epsilon_{AC} - \epsilon}{\epsilon} \approx \lambda \,, \tag{6.79}$$

where Δ_{AC} is an axial conduction[4] effect; ϵ_{AC} is effectiveness for finite axial conduction; and ϵ is effectiveness for no conduction.

Shah [61] gives the following expression for correlation of Bahnke and Howard's [2] data.

$$\begin{aligned} \epsilon_{ROT,AC} &= \epsilon_{CF}\left(1 - \frac{1}{9C_{ROT}^{1.93}}\right)\left(1 - \left[\frac{1}{2 - C_{RAT}}\right] \times \right. \\ &\qquad \left. \left[\frac{1}{1 + NTU - \kappa NTU^3/a^2} - \frac{1}{1 + NTU}\right]\right) \,, \end{aligned} \tag{6.80}$$

[4]In radial-flow regenerators (see Chapters 4 and 5) the relevant direction of conduction is the radial direction. We use the term axial conduction to conform with the literature.

where

$$a \equiv 2\sqrt{NTU^2 + \frac{NTU}{\lambda}} \; ; \tag{6.81}$$

and

$$\kappa \equiv 4 - \left[\frac{8}{a}\right]\left[1 - \frac{1 - e^{-a}}{\sinh a}\right] \; . \tag{6.82}$$

Shah [61] reports that this correlation agrees with Bahnke and Howard's [2] data to within 1% for

1. $1 \leq NTU \leq 20$;
2. $2 \leq C_{ROT} \leq \infty$;
3. $0.9 \leq C_{RAT} \leq 1$;
4. $0.25 \leq (hA)' \leq 1$; and
5. $0 \leq \lambda \leq 0.08$.

From Shah's [61] correlation we can infer the axial-conduction effect given by Equation 6.3.

This section has shown that core conduction in the direction of the fluid flows while the core is exposed to the flows reduces regenerator effectiveness. This section has neglected the effects of core conduction while the core is under seals, however. The next section considers these effects.

6.1.8 Heat Diffusion Under Seals

This section quantifies the effects of core conduction while the core is under seals. This section shows that the effects of this conduction can be large, especially when wide seals are used to limit leakage. As Section 1.2.6 describes, wide seals reduce effectiveness. High-effectiveness regenerators rely on a steep temperature profile through the core. When the core rotates under the seals, the temperature profile degrades toward the mean core temperature. Thus effectiveness is reduced. The degree of reduction depends on the amount of time the core spends under the seals, which is proportional to the width of the seals.

We begin by considering one-dimensional heat diffusion to gain insight into the heat-transfer phenomena that occur while the core is under the seals. Then we consider the effects of seal width on the effectiveness of regenerators.

One-Dimensional Heat Diffusion. To gain insight into the heat-transfer phenomena occurring while a regenerator core is under a seal, we first consider a one-dimensional heat-diffusion problem subject to boundary conditions typical of regenerators. We will show how the temperature profile through a regenerator core degrades toward the mean core temperature while the core is under a seal, and how the degree of degradation is related to the amount of time the core spends under the seal.

A nearly linear temperature profile is usually established in regenerator cores. Therefore, we choose the following representative initial temperature profile.

$$T_R(x, t=0) = mx + b \, , \tag{6.83}$$

where T_R is the core temperature; m is the slope of the linear profile and b is the y-intercept. This yields one boundary condition. The other boundary conditions come from the assumption of no heat transfer from either end of the passage-tubes. The diffusion problem is then

$$\frac{\partial^2 \Theta}{\partial \zeta^2} = \frac{\partial \Theta}{\partial Fo_L} \tag{6.84}$$

where

$$\Theta \equiv \frac{T_R - \overline{T}_R}{T_{Rmax} - \overline{T}_R} \tag{6.85}$$

subject to the following boundary conditions:

1. $\left.\frac{\partial \Theta}{\partial \zeta}\right|_{\zeta=0, Fo_L>0} = 0$

2. $\left.\frac{\partial \Theta}{\partial \zeta}\right|_{\zeta=1, Fo_L>0} = 0$

3. $\Theta(\zeta, Fo_L = 0) = 2\zeta - 1$

The solution to this problem is

$$\Theta(\zeta, Fo_L) = \left[\frac{-8}{\pi^2}\right] \sum_{j=1}^{\infty} \left[\frac{1}{(2j-1)^2}\right] e^{-(2j-1)^2 \pi^2 Fo_L} \cos((2j-1)\pi\zeta) \tag{6.86}$$

Figure 6.8 gives a plot of Equation 6.86. We can see from the plot that the temperature profile flattens at either end of a passage-tube and gradually flattens for the whole passage-tube as the Fourier number (Fo_L) increases. For an actual regenerator, the core passage-tubes would exit from the seals at some Fourier number (Fo_L) with a certain degree of temperature-profile degradation.

Results. The principal author solved the equations given in Section 6.1.4 to quantify the effects of seal width on regenerator effectiveness (see Beck [6]). This section gives his principal results.

We will present the majority of the results given in this section in terms of an axial-conduction-and-seal-width effect,

$$\Delta_{ACS} \equiv \frac{\epsilon_{ACS} - \epsilon}{\epsilon} \, , \tag{6.87}$$

where ϵ_{ACS} is an effectiveness value calculated assuming axial conduction and finite seal widths and ϵ is an effectiveness value calculated assuming no axial conduction and no seal width. This mode of presentation is similar to that

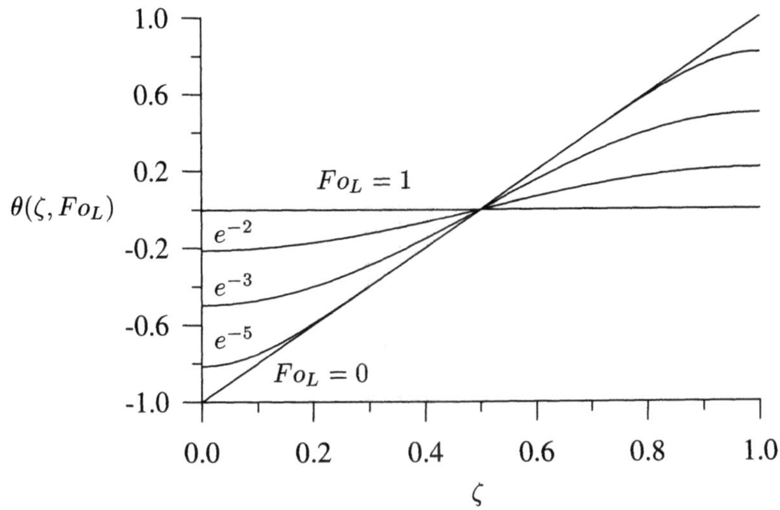

Figure 6.8: Solutions for One-Dimensional Heat Diffusion. Reprinted from Beck [7] with permission from the ASME.

used by Bahnke and Howard (see Bahnke and Howard [2]), except reductions in effectiveness are associated with *negative* axial-conduction-and-seal-width effects in this section, and effectiveness reductions are associated with *positive* axial-conduction effects in Bahnke and Howard's paper. This was done because it was felt that plots of an axial-conduction-and-seal-width effect would be misleading if the effect was defined as positive for reduced effectiveness.

All plots of the axial-conduction-and-seal-width effect (Δ_{ACS}) start at the negative of Bahnke and Howard's axial-conduction effects for a Fourier number of zero ($Fo_L = 0$), which is associated with zero seal width. The Δ_{ACS} values then increase in magnitude as the seal coverages increase and the Fourier numbers (Fo_L) increase. For very large Fourier numbers (for cores almost completely covered with seals), the core-material temperature profiles are completely flat where the passages exit the seals.

Figure 6.9 shows a plot of all of the sample-calculation results that are given in this section. The results are plotted as a seal-width effect vs. the natural logarithm of the Fourier number (Fo_L). We define the seal-width effect as

$$\Delta_S \equiv \frac{\epsilon_{ACS} - \epsilon_{AC}}{\epsilon_{AC} - \epsilon_{ACS}(Fo_L \to \infty)} , \tag{6.88}$$

where ϵ_{AC} is an effectiveness value calculated assuming axial conduction but neglecting seal width (calculated using Bahnke and Howard's [2] method). The effects of seal width have been essentially isolated from the other effects. The plot in Figure 6.9 shows that the effects due to seal width alone are essentially a function of only the Fourier number (Fo_L), increasing in magnitude with

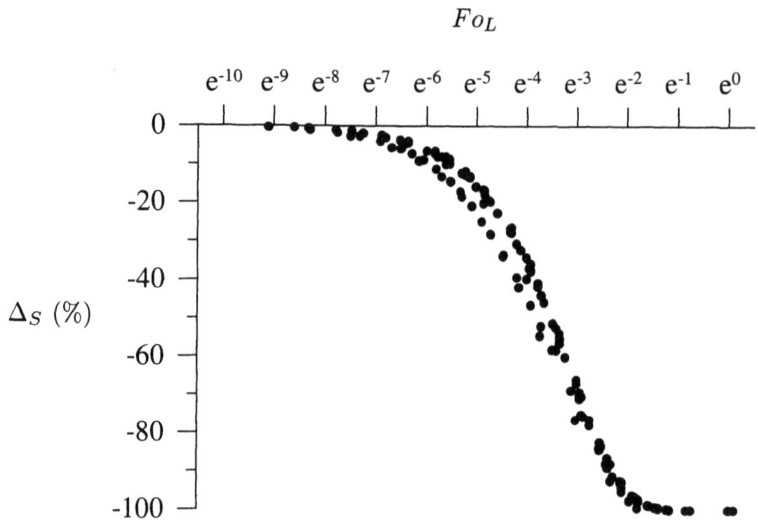

Figure 6.9: Plot of all Sample-Calculation Results Given in this Section.

the Fourier number. This is reasonable, considering the solution to the one-dimensional heat-diffusion problem given above.

The plot in Figure 6.9 may provide insight, but it is not practically useful for three reasons:

1. The plot does not yield design information readily;

2. The plot has too high of an uncertainty to use it for design; and

3. Determining the axial-conduction-and-seal-width effect (ϵ_{ACS}) from the plot would require calculation of the axial-conduction effect (ϵ_{AC}) and the axial-conduction-and-seal-width effect for infinite time ($\epsilon_{ACS}(Fo_L = \infty)$). Each of these calculations take as much computer time as calculation of the axial-conduction-and-seal-width effect (ϵ_{ACS}) directly.

The authors recommend the use of advanced computer codes, such as RGT-OPT™ [5], to calculate regenerator effectiveness directly. the axial-conduction-and-seal-width effect (ϵ_{ACS}) directly. For example, for a regenerator with $NTU = 20$; $C_{RAT} = 0.99$; $C_{ROT} = 3$; $(hA)' = As' = 1$; $\lambda = 0.05$; and $SC = 0.15$, $ln(Fo_L) = -6.5$ and and RGT-OPT calculates $\epsilon_{AC} = 0.9038$; $\epsilon_{ACS} = 0.8923$; and $\epsilon_{ACS}(Fo_L \to \infty) = 0.6036$, so $\Delta_S = -3.983\%$.

To give the designer insight for preliminary design, the following sections give plots of the axial-conduction-and-seal-width effect Δ_{ACS} vs. seal coverage (the fraction of the core's face area occupied by seals) for various values of the heat-transfer parameters. As shown in Fig. 6.9, seal width alone has essentially

[5] RGT-OPT is a trademark of Douglas Stephen Beck, the principal author, and is available from him.

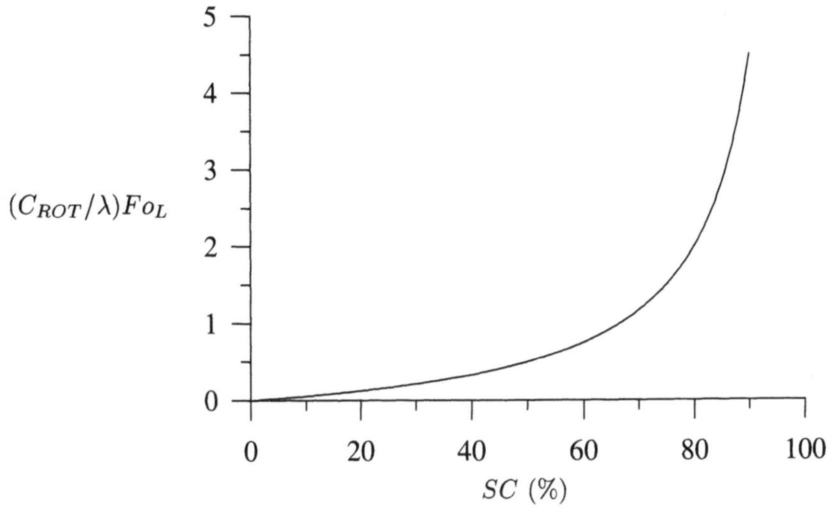

Figure 6.10: Plot of $(C_{ROT}/\lambda)Fo_L$ vs. Seal Coverage.

the same effect on regenerator effectiveness for all values of the heat-transfer parameters. The principal differences between the plots of axial-conduction-and-seal-width effect (Δ_{ACS}) vs. seal coverage are due to

1. The $(C_{ROT}/\lambda)Fo_L$ vs. seal coverage (SC) function (from Equation 1.19) plotted in Figure 6.10 and

2. The dependence of the effects of axial conduction on values of the heat-transfer parameters described by Bahnke and Howard [2].

λ Effects. Figure 6.11 gives plots of the axial-conduction-and-seal-width effect (Δ_{ACS}) vs. seal coverage (SC) for the following values of the heat-transfer parameters: $NTU = 20$; $C_{ROT} = 3$; $C_{RAT} = (hA)' = As' = 1$; and $\lambda = 0.01$, 0.08, and 0.32. Reductions in effectiveness increase in magnitude with the conduction parameter (λ) because the degree of axial conduction increases with λ. All of the plots start at the axial conduction effect (Δ_{AC}) for zero seal coverage. As the core becomes completely covered with seals ($SC \rightarrow 100\%$), the effectivenesses approach those associated with flat temperature profiles where the core exits from under the seals. If the seal coverage is great enough so that the flat-profile condition is reached, the core temperature gradients are small for a large part of the time. This leads to small axial-conduction effects for a large part of the time. This is why the flat-profile effectivenesses are nearly identical.

C_{ROT} Effects. Figure 6.12 shows plots of the axial-conduction-and-seal-width effect (Δ_{ACS}) vs. seal coverage (SC) for the following values of the heat-transfer

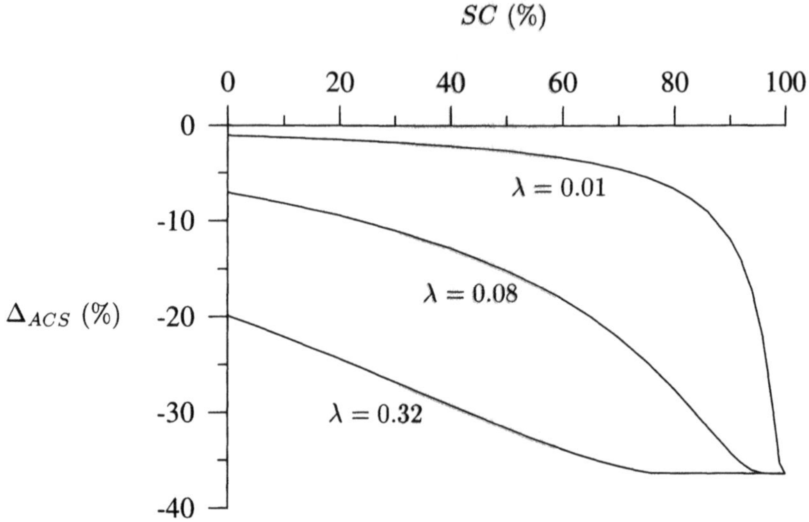

Figure 6.11: Effects of Seal Width on Effectiveness for $NTU = 20$; $C_{ROT} = 3$; $C_{RAT} = (hA)' = As' = 1$; and $\lambda = 0.01$, 0.08, and 0.32. Reprinted from Beck [7] with permission from the ASME.

parameters: $NTU = 20$; $C_{RAT} = (hA)' = As' = 1$; $\lambda = 0.01$; and $C_{ROT} = 1$, 3, and 5. The temperature-profile flattening for high seal coverage is felt most for large values of the dimensionless core-rotation rate (C_{ROT}). For fast core rotation (large C_{ROT}), the core material's temperature profile does not change appreciably for low seal coverage[6]. This leads to high effectiveness. As the seal coverage approaches 100%, however, the flat-profile condition is reached, for which the temperature profile changes drastically. Thus, for high seal coverage, reductions in effectiveness due to axial conduction coupled with finite seal width are greatest for fast core rotation (large C_{ROT}).

***NTU* Effects.** The effect of the dimensionless core size (NTU) on effectiveness reductions is similar to that of the dimensionless core-rotation rate (C_{ROT}). Figure 6.13 shows plots of the axial-conduction-and-seal-width effect (Δ_{ACS}) vs. seal coverage (SC) for the following values of the heat-transfer parameters: $C_{ROT} = 3$; $C_{RAT} = (hA)' = As' = 1$; $\lambda = 0.32$; and $NTU = 1$, 10, and 20. The effect of the dimensionless core size (NTU) on effectiveness reductions is greatest for large NTU. As with the dimensionless core-rotation rate (C_{ROT}), high NTU values produce core-material temperature profiles that do not change appreciably throughout a cycle. Therefore, the flat-profile condition produces significant effectiveness reduction.

[6] For zero seal coverage and infinitely fast core rotation, small variations in the core's axial-temperature profile are indicated by Equations 6.59 and 6.60

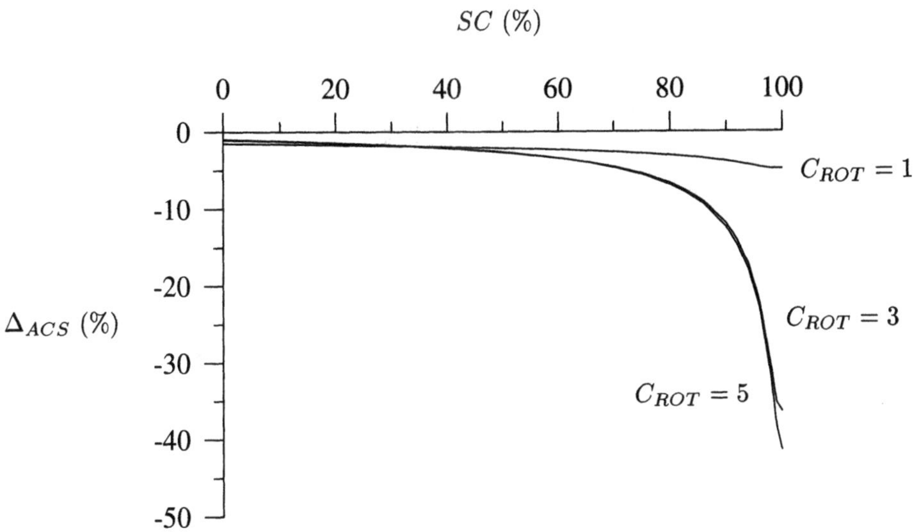

Figure 6.12: Effects of Seal Width on Effectiveness for $NTU = 20$; $C_{RAT} = (hA)' = As' = 1$; $\lambda = 0.01$; and $C_{ROT} = 1$, 3, and 5. Reprinted from Beck [7] with permission from the ASME.

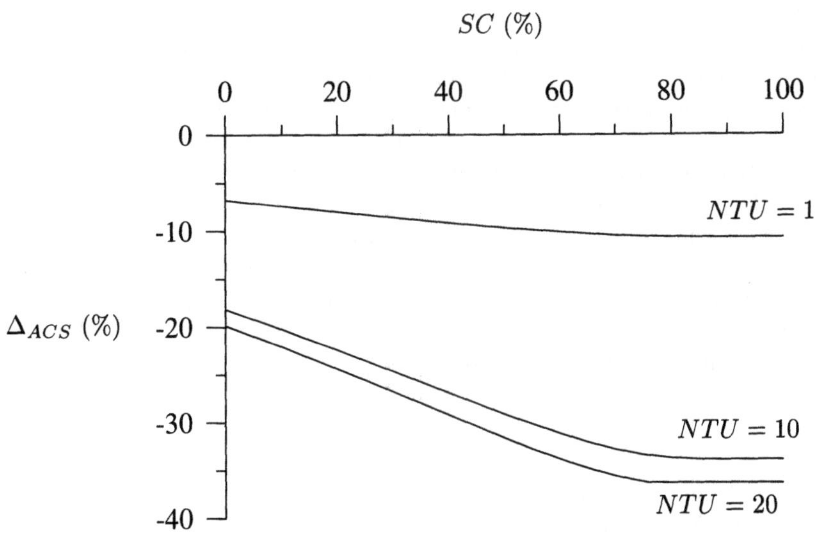

Figure 6.13: Effects of Seal Width on Effectiveness for $C_{ROT} = 3$; $C_{RAT} = (hA)' = As' = 1$; $\lambda = 0.32$; and $NTU = 1$, 10, and 20. Reprinted from Beck [7] with permission from the ASME.

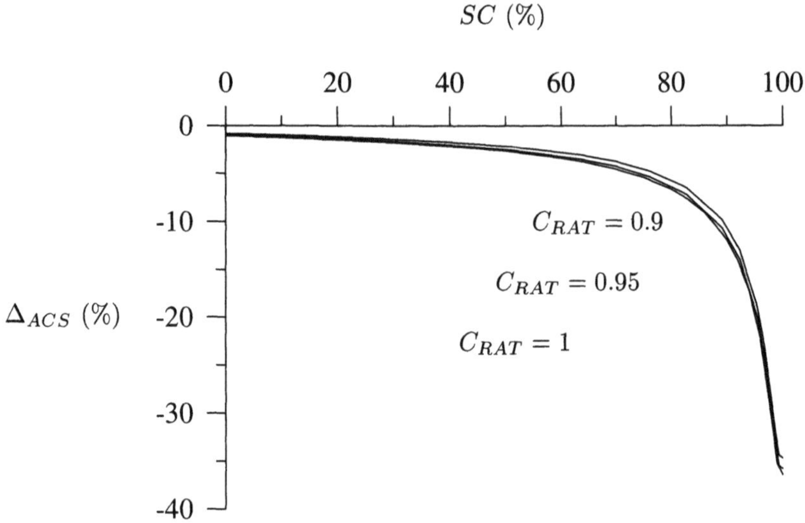

Figure 6.14: Effects of Seal Width on Effectiveness for $NTU = 20$; $C_{ROT} = 3$; $(hA)' = As' = 1$; $\lambda = 0.01$; and $C_{RAT} = 0.90$, 0.95, and 1.0. Reprinted from Beck [7] with permission from the ASME.

C_{RAT} **Effects.** Figure 1.5 shows the effect on effectiveness of varying the heat-capacity-rate ratio (C_{RAT}). As C_{RAT} is decreased (such as by making the exhaust-gas-flow rate much greater than the compressed-air-flow rate for exhaust-gas regenerators) the outlet temperature on the C_X side (exhaust side) increases as does the outlet temperature on the C_N side (compressed-air side). Thus, the effectiveness increases as the heat-capacity-rate ratio (C_{RAT}) is decreased. Axial conduction has little influence. Bahnke and Howard [2] report little variation in axial-conduction effect with the heat-capacity-rate rate (C_{RAT}). We expect little variation in axial-conduction-and-seal-width effect with C_{RAT} because this effect depends on axial conduction. The plots of axial-conduction-and-seal-width effect (Δ_{ACS}) vs. seal coverage (SC) shown in Figure 6.14 demonstrate that this is the case.

$(hA)'$ **and** As' **Effects.** Bahnke and Howard [2] found little variation in axial-conduction effect (Δ_{AC}) when As' differed from $(hA)'$ by $\pm 30\%$. Also, they found little variation in axial-conduction effect with simultaneous variation of As' and $(hA)'$. Beck [6, 7] found small variation of effectiveness reductions when finite axial conduction coupled with finite seal width is considered. Figure 6.15 shows plots of the axial-conduction-and-seal-width effect (Δ_{ACS}) vs. seal coverage (SC) for $NTU = 20$; $C_{ROT} = 3$; $C_{RAT} = 1.0$; $\lambda = 0.32$; and $(hA)' = As' = 0.25$ and 1.0.

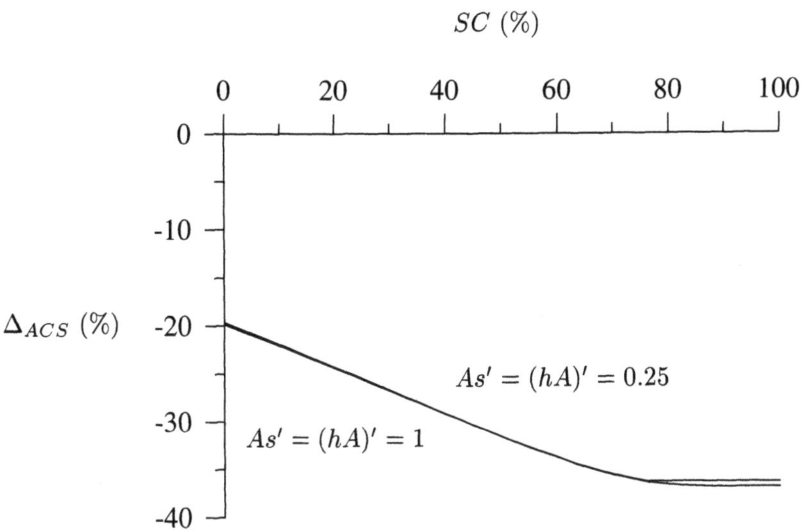

Figure 6.15: Effects of Seal Width on Effectiveness for $NTU = 20$; $C_{ROT} = 3$; $C_{RAT} = 1.0$; $\lambda = 0.32$; and $(hA)' = As' = 0.25$ and 1.0. Reprinted from Beck [7] with permission from the ASME.

Example Calculation. Now we consider an example regenerator to show that regenerator effectiveness can be greatly reduced due to seal width.

Consider the modular regenerator shown in Figure 6.16 (see Chapter 4 for a detailed description of the modular regenerator). The seal coverage of this regenerator (the fraction of the modules not exposed to flows) is

$$SC \quad = \quad 5/15 \quad = \quad 33.3\% \, . \tag{6.89}$$

The solid-area ratio is $As' = 1$. Assume that the solid area on the compressed-air side is $As_N = 2m^2$; the thermal conductivity of the core material is $k_R = 0.2W/m{-}K$; the heat-capacity rate of the compressed-air flow is $C_N = (\dot{m}c_p)_N = (0.25kg/s)(1,050J/kg - K)$; and the core thickness is $L = 0.038m$. Then the dimensionless conduction parameter is

$$\lambda \quad = \quad \tfrac{(0.2)(2)}{(0.038)(0.25)(1050)}\left(1 + \tfrac{1}{1}\right) \quad = \quad 0.08 \, . \tag{6.90}$$

From Figure 6.11, the axial-conduction-and-seal-width effect is $\Delta_{ACS} \approx -12\%$ and the axial-conduction effect is $\Delta_{AC} \approx -7.5\%$. Thus, seal width contributes greatly to effectiveness reduction due to core conduction.

Summary of Seal-Width Effects. The following summarizes the effects of seal width on regenerator effectiveness.

1. Seal width should be considered when one considers axial-conduction effects: reductions in effectiveness due to axial conduction coupled with

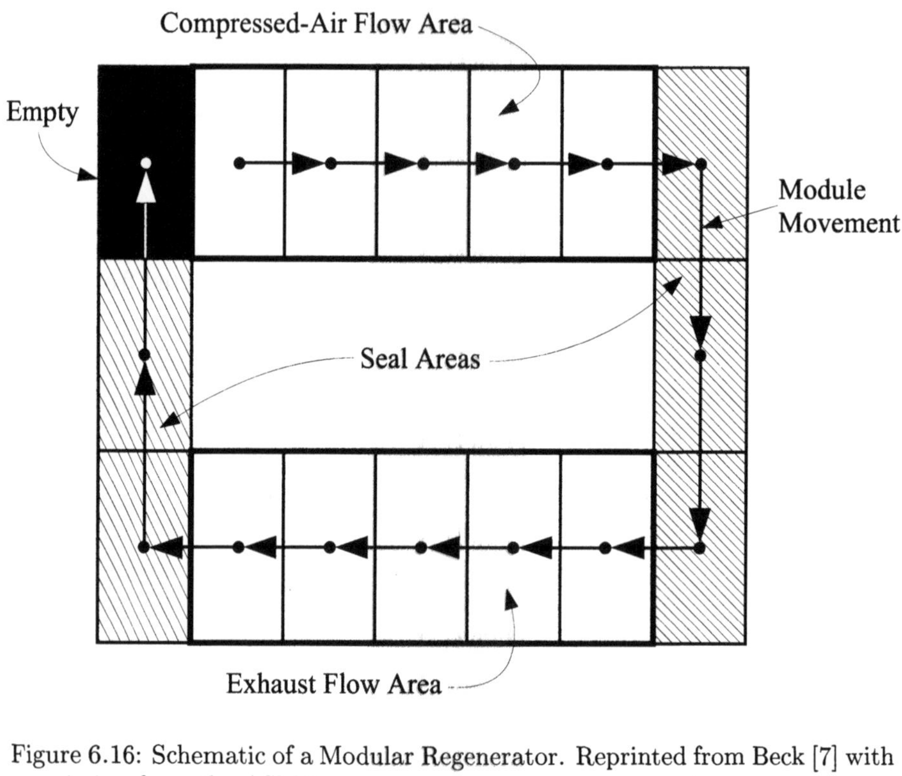

Figure 6.16: Schematic of a Modular Regenerator. Reprinted from Beck [7] with permission from the ASME.

finite seal width can be twice as great as those due to axial conduction alone.

2. The effects due to axial conduction coupled with finite seal width vary greatly with the core-conduction parameter (λ) and the dimensionless core size (NTU), being greatest for high core conduction (large λ) and/or large cores (large NTU).

3. The effects differ to a lesser degree with the dimensionless core-rotation rate (C_{ROT}).

4. The effects differ to a small degree with the heat-capacity-rate ratio (C_{RAT}).

5. The effects differ negligibly for variations in the solid-area ratio (As') and the convective-conductance ratio ($(hA)'$) within limits usually experienced in gas-turbine applications.

6.1.9 Seal-Width Effect

Finite-difference effectiveness calculations can take a significant amount of time (about thirty seconds with a 486 computer). The principal author used data calculated using the code listed in Reference [7] to obtain the simple expression for the seal-width effect given by Equation 6.6. This expression can be used to rapidly obtain effectiveness values that are within 1% accuracy for typical gas-turbine regenerators.

6.1.10 Porosity

This section quantifies the effects of low porosity on heat-transfer performance. This section follows the work of Beck [6]. First, we give the key results of Beck's [6] analyses of the effects of low porosity. Then, we list finite-difference equations that Beck [6] used in his analyses. Finally we give other results and conclusions.

Key Results. Figure 6.17 shows the key effects of low porosity on the heat-transfer performance of regenerators. Low porosities are associated with thick passage walls, into which it is difficult to transfer heat to the central portions due to the relatively high thermal resistance. Thus, heat-transfer performance is reduced for thick passage walls and low porosities. Section 1.2.7 defined a porosity number, N_p (see Equation 1.24), which is used to quantify the effects of low porosity. Figure 6.17 shows the effects of low porosity in terms of the porosity number and for two specific cases: (1) for ceramic cores and (2) for stainless-steel cores. Figure 6.17 shows that heat-transfer performance is reduced to the greatest extent in ceramic cores, for which the heat-transfer resistance is reduced by about 12% at a porosity of $p = 0.5$. The effects of low porosity for stainless-steel cores are negligible for all but very low porosities due to the relatively high thermal conductivity of stainless steel.

The following describes the effects of low porosity in detail.

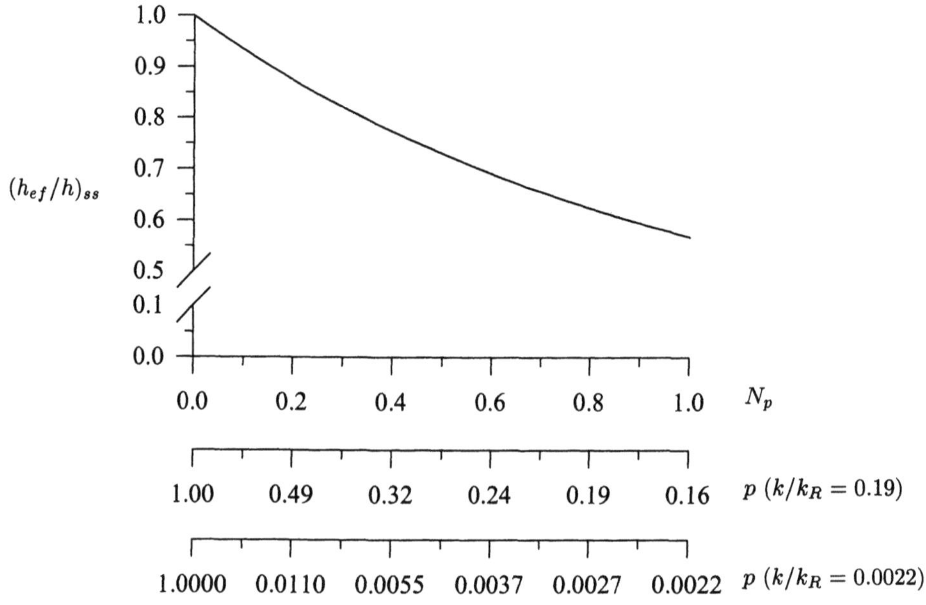

Figure 6.17: Effects of Low Porosity on the Heat-Transfer Performance of Regenerators (lower scales are for ceramic cores $[k/k_R = 0.19]$ and stainless-steel cores $[k/k_R = 0.0022]$).

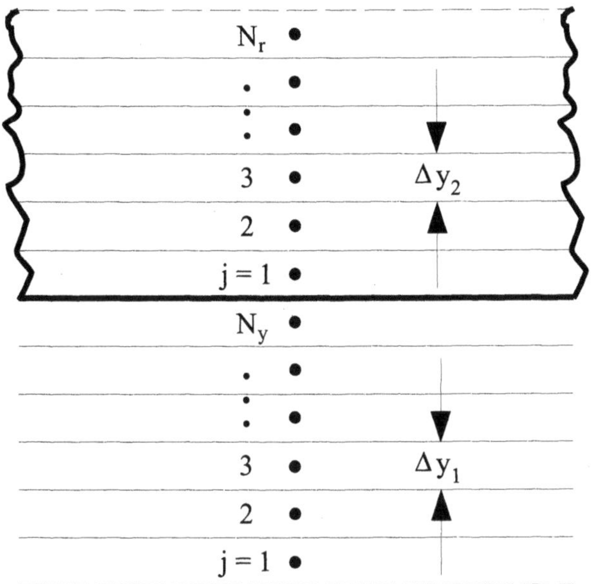

Figure 6.18: Isolated Half-Passage-Tube of a Parallel-Plate Regenerator.

Finite-Difference Equations. We start by deriving the finite-difference equations that Beck [6] used for his analyses of the effects of low porosity on heat-transfer performance. Section 6.1.4 gives the coupled energy equations for the core-material and fluid-flow domains. No solution to these equations was attempted. Instead, Beck [6] approximated these equations with a set of finite-difference equations, and he used a computer program to generate the results given below.

Consider the schematic of an isolated half-passage-tube as shown in Figure 6.18. The fluid domain has been divided into N_y sections, each of height $\Delta y_1 = y_0/N_y$. Similarly, the core-material domain has been divided into N_R sections, each of height $\Delta y_2 = H/N_R$.

Finite-Difference Equations for the Momentum Equation

From Equation 6.45, the normalized velocities for each fluid-domain section are

$$u_j^\star = \frac{3N_y^2 - 3j^2 + 3j - 1}{2N_y^2}$$

$$1 \le j \le N_y \quad . \tag{6.91}$$

Finite-Difference Equations for the Energy Equation in the Fluid Domain

Consider the following Taylor expansions about points in the fluid domain. If $j = 1$,

$$\Theta(y_1 + \Delta y_1) \approx \Theta(y_1) + \left.\frac{d\Theta}{dy_1}\right|_{y_1} \Delta y_1 + \left.\frac{d^2\Theta}{dy_1^2}\right|_{y_1} \frac{(\Delta y_1)^2}{2} \tag{6.92}$$

$$\Theta(y_1 - \Delta y_1) \approx \Theta(y_1) - \left.\frac{d\Theta}{dy_1}\right|_{y_1} \Delta y_1 + \left.\frac{d^2\Theta}{dy_1^2}\right|_{y_1} \frac{(\Delta y_1)^2}{2} , \tag{6.93}$$

where

$$\Theta \equiv \frac{T_0 - T_1}{T_0 - T_m} . \tag{6.94}$$

From symmetry at the center line,

$$\Theta(y_1 - \Delta y_1) = \Theta(y_1) . \tag{6.95}$$

Therefore,

$$\left.\frac{d^2\Theta}{dy_1^2}\right|_{y_1} \approx \frac{\Theta(y_1 + \Delta y_1) - \Theta(y_1)}{(\Delta y_1)^2} , \tag{6.96}$$

or

$$\frac{d^2\Theta}{dy_1^{\star 2}} \approx N_y^2 (\Theta_2 - \Theta_1) . \tag{6.97}$$

Similar calculations can be performed for the other sections to yield the following relations. If $j = N_y$,

$$\frac{d^2\Theta}{dy_1^{\star 2}} \approx \left(\frac{4N_y^2}{3}\right) (\Theta_{N_y - 1} - 3\Theta_{N_y}) \tag{6.98}$$

Otherwise,

$$\frac{d^2\Theta}{dy_1^{\star 2}} \approx N_y^2 (\Theta_{j+1} + \Theta_{j-1} - 2\Theta_j) ,$$

$$1 < j < N_y . \tag{6.99}$$

From Equations 6.52 and 6.97–6.99, we construct the following vector equation:

$$\mathbf{A}_1 \mathbf{x}_1 = \left(-\frac{Nu_H}{4}\right) \mathbf{b} , \tag{6.100}$$

where $\mathbf{x}_1 \in \Re^{N_y}$; $\mathbf{b} \in \Re^{N_y}$; $x_{1_j} = \Theta_j$; and $b_j = u_j^{\star}$. Also, from the definitions of a mixed-mean temperature and the dimensionless temperature (Θ),

$$\mathbf{b}^T \mathbf{x}_1 = N_y . \tag{6.101}$$

Equations 6.100 and 6.101 are solved in three steps. First, solve

$$\mathbf{A}_1 \left[\left(\frac{1}{Nu_H} \right) \mathbf{x}_1 \right] = \left(-\frac{1}{4} \right) \mathbf{b} \; . \tag{6.102}$$

Then, solve

$$Nu_H = \frac{N_y}{\mathbf{b}^T \left[\left(\frac{1}{Nu} \right) \mathbf{x}_1 \right]} \; . \tag{6.103}$$

Finally,

$$\mathbf{x}_1 = Nu_H \left[\left(\frac{1}{Nu_H} \right) \mathbf{x}_1 \right] \; . \tag{6.104}$$

Using this three-step procedure, we can calculate both the Nusselt number and the dimensionless temperature profile in the fluid domain.

Equations for the Energy Equation in the Core-Material Domain

In the core-material domain, similar approximations for the relevant derivatives as were made above can be made for the point at the center of the wall and interior points. Namely, for $j = N_R$,

$$\frac{\partial^2 T_2}{\partial y_2^{\star 2}} \approx N_R^2 \left(T_{2_{N_R - 1}} - T_{2_{N_R}} \right) \; , \tag{6.105}$$

and

$$\frac{\partial^2 T_2}{\partial y_2^{\star 2}} \approx N_R^2 \left(T_{2_{j+1}} + T_{2_{j-1}} - 2T_{2_j} \right) \; ,$$

$$1 < j < N_R \; . \tag{6.106}$$

For the section bordering the fluid/solid interface,

$$\frac{\partial^2 T_2}{\partial y_2^{\star 2}} \approx \left(\frac{4N_R^2}{3} \right) \left(T_{2_2} + 2T_0 - 3T_{2_1} \right) \; . \tag{6.107}$$

The wall temperature can be evaluated by equating the heat flux at the wall in the fluid domain with the heat flux at the wall in the core-material domain. Doing so,

$$T_0 = \left(\frac{\Delta}{4 + \Delta} \right) T_m + \left(\frac{1}{4 + \Delta} \right) T_{2_2} + \left(\frac{3}{4 + \Delta} \right) T_{2_1} \; , \tag{6.108}$$

where

$$\Delta \equiv N_p \left(\frac{N_y}{N_R} \right) \left(\Theta_{N_y - 1} + 3\Theta_{N_y} \right) \; . \tag{6.109}$$

Thus, for $j = 1$,

$$\frac{\partial^2 T_2}{\partial y_2^{\star 2}} \approx \left[\frac{4N_R^2}{3} \right] \left[\left(\frac{6 + \Delta}{4 + \Delta} \right) T_{2_2} + \left(\frac{2\Delta}{4 + \Delta} \right) T_m - \left(\frac{6 + 3\Delta}{4 + \Delta} \right) T_{2_1} \right] \; . \tag{6.110}$$

From Equations 6.54, 6.105, 6.106, and 6.110, the following dynamic vector equation is obtained.

$$\dot{\mathbf{x}}_2(Fo_H) = \mathbf{A}_2\mathbf{x}_2(Fo_H) \ , \tag{6.111}$$

where $\mathbf{x}_2 \in \Re^{N_R}$; and $x_{2_j} = T_{2_j}$. Note that the porosity number (N_p) is the only parameter on which the temperature dynamics in the core-material domain depend. Thus, the effects of low porosity on heat-transfer performance are a function of only the porosity number and the Fourier number (and only the porosity number at steady state).

Discussion of Results. Beck [6] solved the finite-difference equations given in the last section using Matlab[7].

The portion of a core first entering one of the flows has an initial axial temperature profile that is nearly linear. As this core portion rotates through the flow, the mixed-mean fluid-flow temperature at any axial location varies (for a passage exposed to the hot flow, the flow temperature increases, while for a passage exposed to the cold flow, the flow temperature decreases). First, we consider a passage cross section with a constant initial temperature profile of 0° that is exposed to a constant flow temperature of 1°. Then we consider fluid-flow-temperature dynamics. Finally we consider initial temperature profiles.

Passages with a Constant Initial Temperature Profile Exposed to a Constant Flow Temperature

Figures 6.19 and 6.20 show plots of core-material temperature profiles for a porosity number of $N_p = 0.1$. For small time ($Fo_H < 0.1$), the temperature of the fluid/solid interface (T_0) rises quickly while the average wall temperature ($\overline{T}_w$) lags behind as the heat diffuses into the core material. For $0.1 < Fo_H < 1.0$, the profiles become more linear as the heat propagates into the core material. Also, the slopes at the interface flatten as the surface temperature increases toward the interface temperature ($T_0 = 1$) and less heat is transfered to the wall due to the smaller temperature difference.

Figures 6.21 and 6.22 show plots of core-material temperature profiles for a smaller porosity number: $N_p = 0.01$. These plots are similar in shape to the plots for $N_p = 0.1$. The temperatures are smaller for all time (or Fourier number, Fo_H), however. This is because a lower porosity number is associated with a higher porosity (a thinner wall for a given hydraulic diameter, D_H) for a given thermal-conductivity ratio (k/k_R). Alternatively, a lower porosity number is associated with a lower conductivity ratio for a given porosity. If the porosity is increased the wall becomes thinner, and a given Fourier number (Fo_H) is associated with a smaller time. If the conductivity ratio (k/k_R) is decreased, the conductivity of the core material and also the diffusivity of the core material increase, and a given Fourier number again is associated with a smaller time. Each temperature profile in Figures 6.21 and 6.22 has been effected more quickly than the profile associated with an identical Fourier number in Figures 6.19

[7]©Copyright The MathWorks, Inc. 1984–1992.

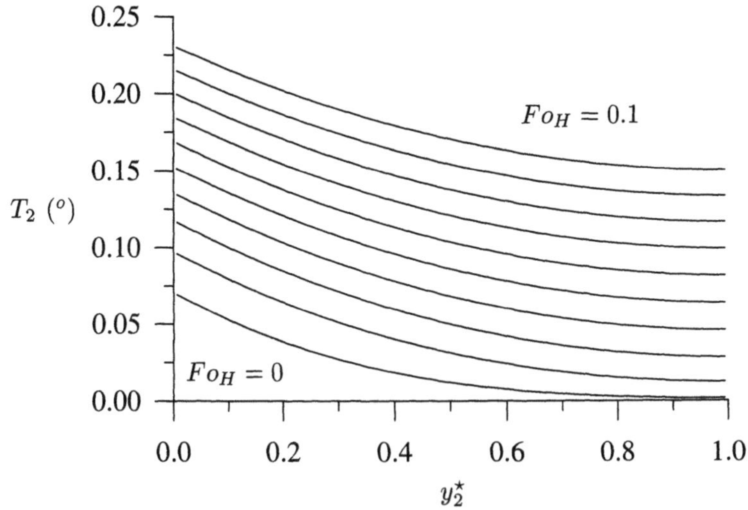

Figure 6.19: Core-Material Temperature Profiles for $N_p = 0.1$ and $0 \leq Fo_H \leq 0.1$.

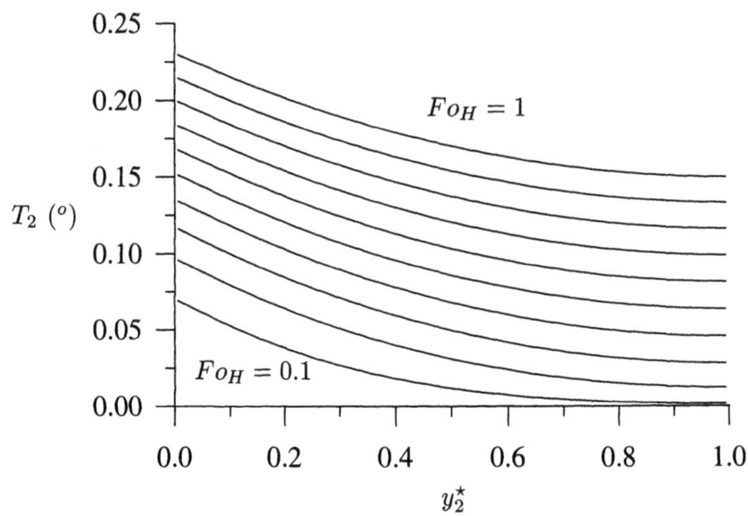

Figure 6.20: Core-Material Temperature Profiles for $N_p = 0.1$ and $0.1 \leq Fo_H \leq 1$.

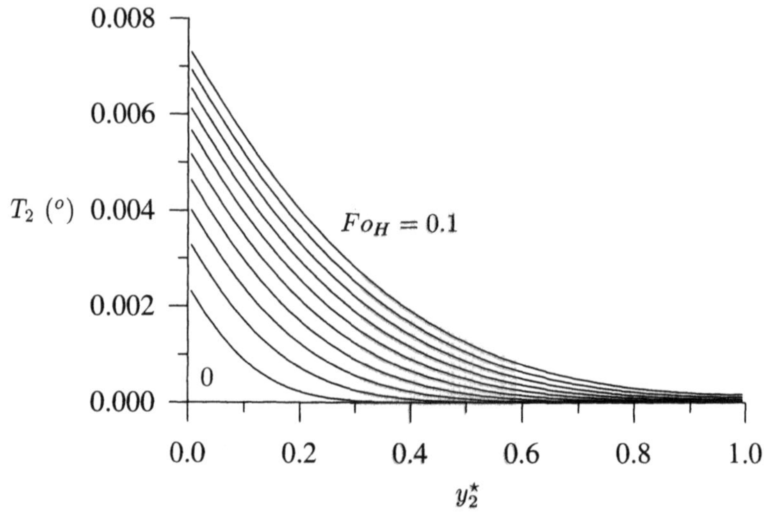

Figure 6.21: Core-Material Temperature Profiles for $N_p = 0.01$ and $0 \leq Fo_H \leq 0.1$.

and 6.20. Therefore, the temperatures for identical Fourier numbers are lower in Figures 6.21 and 6.22 than in Figures 6.19 and 6.20.

Figures 6.23 and 6.24 show plots of the heat-transfer-coefficient ratio (h_{ef}/h) vs. Fourier number (Fo_H) for $N_p = 0.1$ and $N_p = 0.01$, respectively. Each plot begins at 1 for zero time ($Fo_H = 0$). The interface temperature (T_0) and the mean wall temperature ($\overline{T}_w$) are both equal to 0° for zero time. Therefore, the ratio of temperature differences given by Equation 1.23 is equal to 1 for $Fo_H = 0$. For small time ($Fo_H < 0.1$), the interface temperature (T_0) rises quickly while the average wall temperature ($\overline{T}_w$) lags behind, as shown in Figures 6.19 and 6.21. This reduces the ratio of temperature differences, which reduces h_{ef}/h. For larger time ($Fo_H > 0.1$), a steady-state h_{ef}/h value, $(h_{ef}/h)_{ss}$, is approached. This steady-state value decreases as the porosity number (N_p) is increased. This means that small reductions in heat-transfer performance due to porosity can be achieved with a high-porosity core or a core made of a highly conductive material.

The trends described above occur for sufficiently low porosity number. For higher porosity number (low p or high k/k_r), the interface temperature (T_0) is limited by $1^o = T_m$, producing the temperature-profile plots shown in Figure 6.25 for $N_p = 1$, which should be compared with the plots in Figures 6.20 and 6.22. Limiting the interface temperature for high porosity number allows the wall temperature ($\overline{T}_w$) to catch up, bringing h_{ef}/h to a steady-state value more quickly in terms of the Fourier number (in terms of time, temperature responses are slower for high porosity number than for low N_p).

To compare responses of the heat-transfer-coefficient ratio (h_{ef}/h) in terms

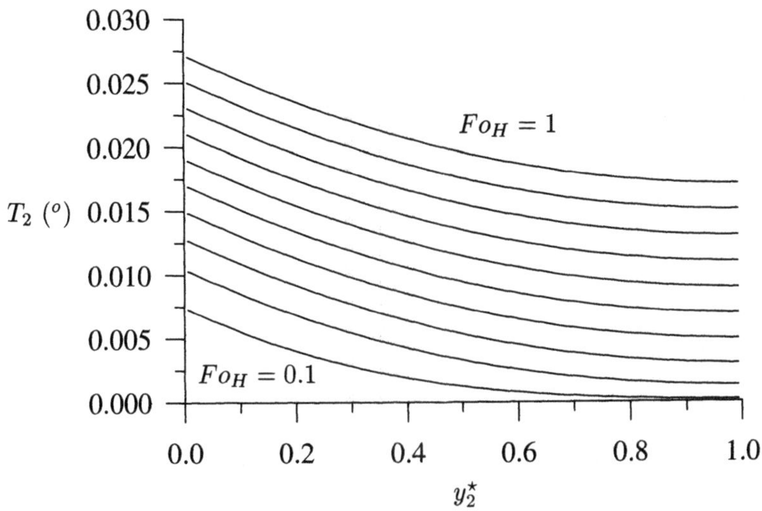

Figure 6.22: Core-Material Temperature Profiles for $N_p = 0.01$ and $0.1 \leq Fo_H \leq 1$.

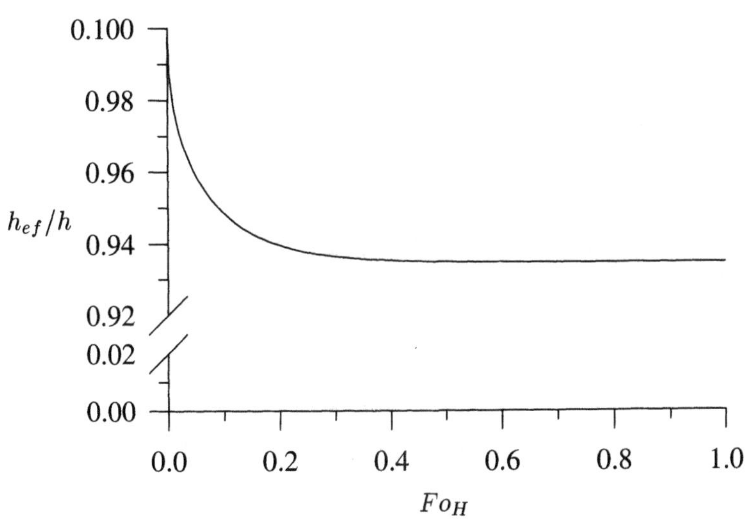

Figure 6.23: Reduction in Core Heat-Transfer Performance vs. Dimensionless Time for $N_p = 0.1$.

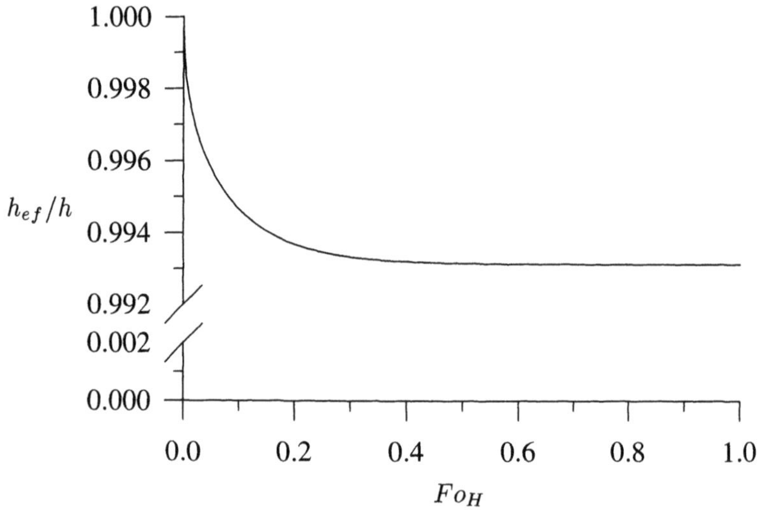

Figure 6.24: Reduction in Core Heat-Transfer Performance vs. Dimensionless Time for $N_p = 0.01$.

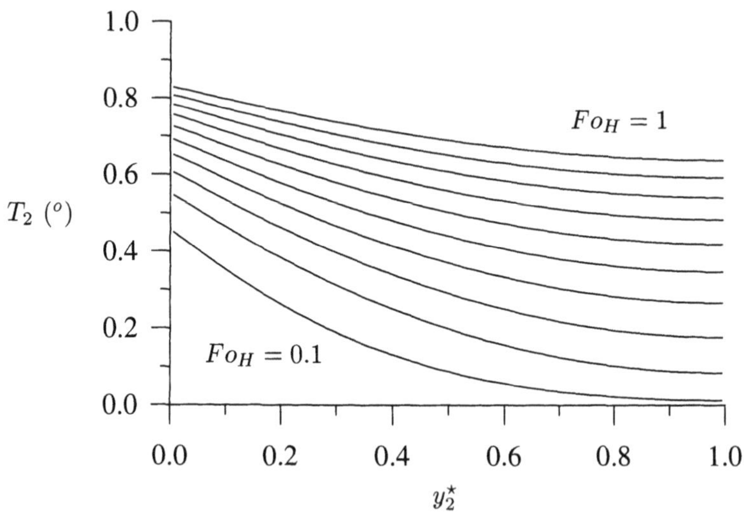

Figure 6.25: Core-Material Temperature Profiles for $N_p = 1$ and $0.1 \leq Fo_H \leq 1$.

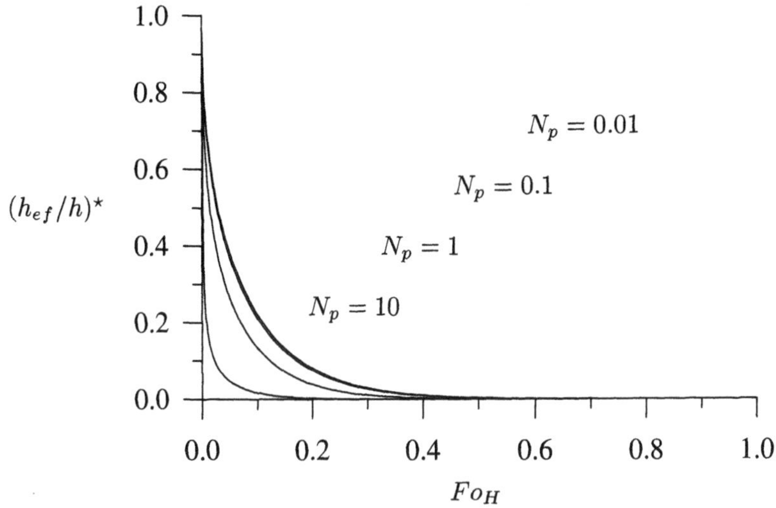

Figure 6.26: Normalized Reductions in Core Heat-Transfer Performance vs. Dimensionless Time for Various N_p.

of Fourier number (Fo_H) for different porosity numbers (N_p), we normalize h_{ef}/h:

$$\left(\frac{h_{ef}}{h}\right)^\star \equiv \frac{h_{ef}/h - (h_{ef}/h)_{ss}}{1 - (h_{ef}/h)_{ss}} \,. \qquad (6.112)$$

Figure 6.26 shows plots of $(h_{ef}/h)^\star$ vs. Fo_H for various N_p. For sufficiently low porosity number the responses are nearly identical, while for higher porosity number the responses are quicker in terms of the Fourier number.

Fluid-Flow-Temperature Dynamics

In the previous sections we considered parallel-plate passage cross sections with initial temperatures of 0° exposed to a fluid flow with a constant mixed-mean temperature of $T_m = 1^o$. This section considers a temporally varying flow temperature.

The core portion that is exposed to the most dramatic flow-temperature dynamics on the exhaust side (C_X side) is physically situated immediately before the outlet. Figure 6.27 shows the exhaust-side outlet-temperature response of a regenerator with $NTU = 10$; $C_{ROT} = 0.35$; $C_{RAT} = 1.0$; $(hA)' = 1.0$; $As' = 1.0$; $\lambda = 0.01$; and $SC = 0.2$ (these system-parameter values produce an effectiveness of $\epsilon = 0.35$ for $p \to 1$). This figure also shows a plot of the model temperature response given by

$$T_{Xo}(Fo_H) = \begin{cases} 0.5\left[1 - \cos\left(\pi \left\{Fo_H/Fo_{Hf}\right\}\right)\right] & \text{if } Fo_H \leq Fo_{Hf} \\ 1 & \text{if } Fo_H > Fo_{Hf} \end{cases} \,, \qquad (6.113)$$

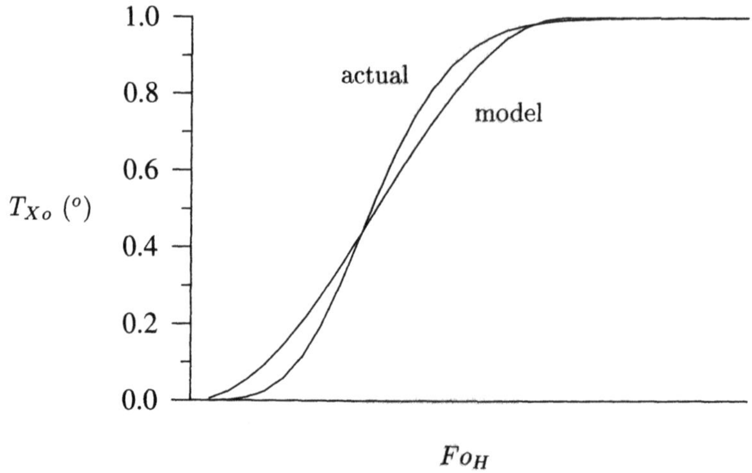

Figure 6.27: C_X-Side Outlet-Temperature Response of a Regenerator with $NTU = 10$; $C_{ROT} = 0.35$; $C_{RAT} = 1.0$; $(hA)' = 1.0$; $As' = 1.0$; $\lambda = 0.01$; and $SC = 0.2$ and a Model Temperature Response.

for some value of a final Fourier number, Fo_{Hf}.

Consider the effects of varying the final Fourier number (Fo_{Hf}). If Fo_{Hf} is sufficiently small (much smaller than $Fo_H = 1$ for example) then core-material temperature responses and h_{ef}/h responses should not be much different than the responses given above. On the other hand, if Fo_{Hf} is much larger than $Fo_H = 1$, then the core-material temperature profile should come to some steady-state during which it follows the changing flow temperature until the flow temperature levels-off at a value of $T_m = 1^o$ at $Fo_H = Fo_{Hf}$.

Figure 6.28 shows plots of the heat-transfer-coefficient ratio (h_{ef}/h) vs. Fourier number (Fo_H) for $N_p = 0.1$ and various final Fourier numbers (Fo_{Hf}). For low Fo_{Hf}, the h_{ef}/h responses are similar to the response in Figure 6.23. As Fo_{Hf} increases, however, the plots level-off at some value (which is greater than the $(h_{ef}/h)_{ss}$ value in Figure 6.23) until $Fo_H \approx Fo_{Hf}$. At $Fo_H \approx Fo_{Hf}$, h_{ef}/h falls to $(h_{ef}/h)_{ss}$.

In Figure 6.28, the difference between h_{ef}/h and $(h_{ef}/h)_{ss}$ that is incurred for high Fo_{Hf} and $5 < Fo_H < Fo_{Hf}$ is small. Beck [6] observed that this difference increases with N_p, but remains a small percentage of the total difference between 1 and $(h_{ef}/h)_{ss}$. Therefore, the effects of fluid-flow-temperature dynamics can be neglected in most analyses.

Initial Temperature Profile

The previous sections have considered a constant initial temperature profile. Consider the effects of an arbitrary initial temperature profile. In a regenerator,

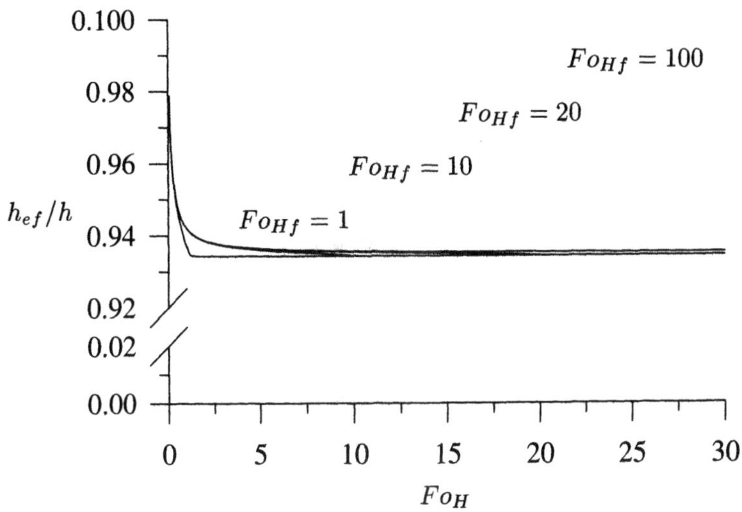

Figure 6.28: Reductions in Core Heat-Transfer Performance for Various Flow-Temperature Dynamics.

the initial temperature profile that would alter the results the most would be obtained with very thin seals. Thick seals allow the heat in the wall to diffuse, which flattens the temperature profile through the wall.

Figures 6.29 and 6.30 show temperature profiles in a core wall with $N_p = 0.1$. The passage was exposed to a hot flow with $T_m = 1^o$ for $Fo_H = 1$ and then a cold flow with $T_m = 0^o$. The profiles plotted are for the passage's cold-flow exposure. For Fourier numbers in the range $0.1 < Fo_H < 1$, the only plot that is affected by the spatially varying initial temperature profile is the plot for $Fo_H = 0.1$ (Compare Figure 6.30 with Figure 6.20).

Figure 6.31 shows a plot of the heat-transfer-coefficient ratio (h_{ef}/h) for the same conditions. This plot starts at an h_{ef}/h value that is greater than one because the mixed-mean temperature is zero: $T_m = 0^o$. However, the heat-transfer-coefficient ratio falls to the same steady-state value as for the plot in Figure 6.23 in a similar Fourier number (Fo_H). We conclude that a spatially varying initial temperature profile of the core material has little effect on the response time of the heat-transfer-coefficient ratio (h_{ef}/h) and on its steady-state value $[(h_{ef}/h)_{ss}]$.

Criteria for Use of a Constant Heat-Transfer-Coefficient Ratio $[(h_{ef}/h)_{ss}]$

The previous sections gave plots showing reductions in heat-transfer performance due to low porosity (plots of h_{ef}/h vs. Fo_H) for parallel-plate-regenerator passages[8]. It is of interest to determine the conditions under which the constant

[8] As discussed in Section 1.2.7, effects of low porosity on heat-transfer performance are relevant for only parallel-plate regenerators

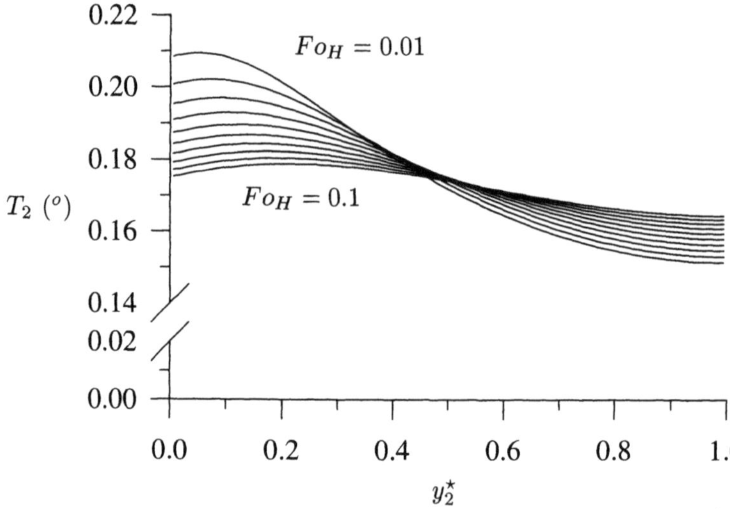

Figure 6.29: Temperature Profiles for $Fo_H < 0.1$ in a Core Wall with $N_p = 0.1$ and a Spatially Varying Initial Temperature Profile.

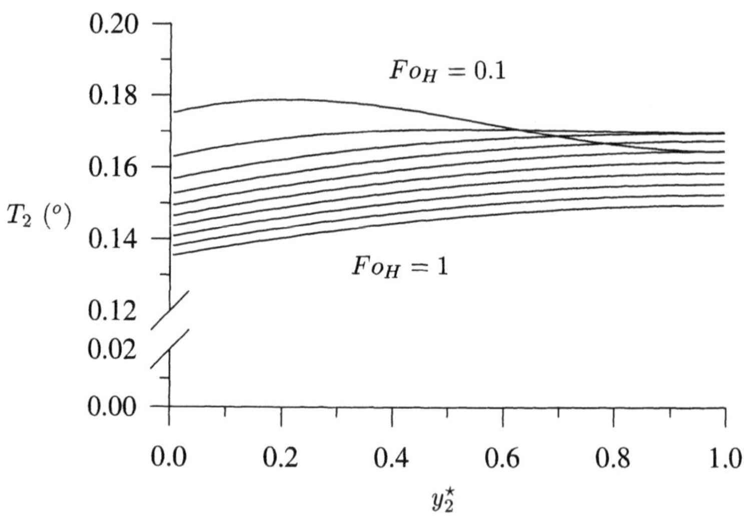

Figure 6.30: Temperature Profiles for $0.1 < Fo_H < 1$ in a Core Wall with $N_p = 0.1$ and a Spatially Varying Initial Temperature Profile.

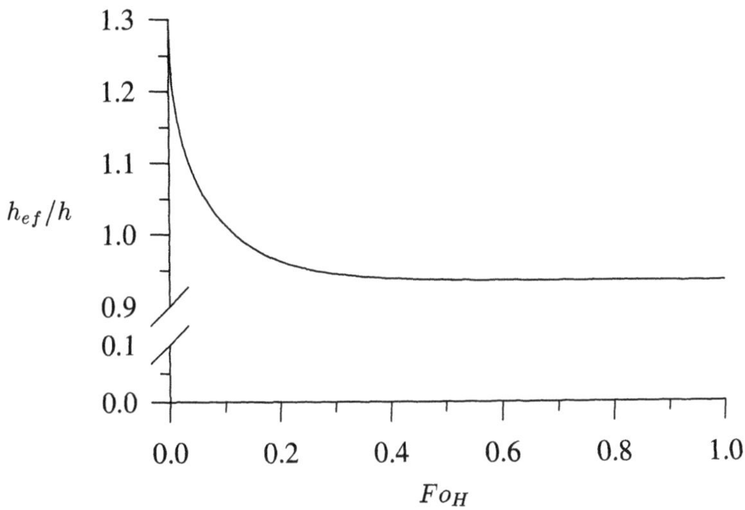

Figure 6.31: Reductions in Core Heat-Transfer Performance for a Core with a Spatially Varying Initial Temperature Profile and $N_p = 0.1$.

steady-state value for the heat-transfer-coefficient ratio $[(h_{ef}/h)_{ss}]$ value can be used to analyze regenerator performance. Use of the constant value would simplify the analysis.

Section 6.1.10, showed that fluid-flow-temperature dynamics can extend the response time of the heat-transfer-coefficient ratio (h_{ef}/h) to large Fourier numbers $(Fo_H \approx 5)$, but the differences between h_{ef}/h and $(h_{ef}/h)_{ss}$ are small for these large Fourier numbers. Therefore, negligible error should be incurred in regenerator-performance calculations if a constant h_{ef}/h value equal to $(h_{ef}/h)_{ss}$ is assumed for passage exposure times associated with sufficiently high Fourier numbers (e.g. $Fo_H > 30$). Thus, a criterion for the compressed-air side (C_N side) is

$$\frac{\alpha_R \tau_N}{H^2} \gg 1 . \tag{6.114}$$

The exposure time to the compressed air can be expressed as

$$\tau_N = \frac{4H^2 NTU[1 + (hA)']}{N_p C_{ROT} Nu_H \alpha_R} . \tag{6.115}$$

Therefore, the criterion for use of a constant heat-transfer-coefficient ratio $[(h_{ef}/h)_{ss}]$ on the compressed-air side in regenerator performance calculations is

$$\frac{4NTU[1 + (hA)']}{N_p C_{ROT} Nu_H} \gg 1 . \tag{6.116}$$

The analogous criterion for use of $(h_{ef}/h)_{ss}$ on the exhaust side (C_X side) in regenerator-performance calculations is

$$\left(\frac{1}{As'}\right)\frac{4NTU[1+(hA)']}{N_p C_{ROT} Nu_H} \gg 1 \,. \qquad (6.117)$$

Consider the following values for regenerator system parameters:

$$\begin{aligned}
\frac{k}{k_R} &= 0.19\,;\\
p &= 0.5\,;\\
NTU &= 20\,;\\
(hA)' &= 1\,;\\
C_{ROT} &= 3\,;\\
Nu_H &= 8.235\,;\ \text{and}\\
As' &= 1\,,\ \text{for which}\\
\epsilon &\approx 0.94\,.
\end{aligned}$$

The value $k/k_R = 0.19$ would be typical for a ceramic regenerator. Current regenerators have porosities of $p \approx 0.7$–0.8. Therefore, $p = 0.5$ is conservative in satisfying the criteria given above. The value $NTU = 20$ (considered with the other system parameters) is roughly associated with a regenerator of $\epsilon = 0.94$, a high-effectiveness regenerator. The Nusselt number for the parallel-plate geometry with a constant wall heat flux is $Nu_H = 8.235$. Hagler [23] found that $C_{ROT} = 3$ is optimal for the regenerative gas turbine that she considered. We found an optimal value of $C_{ROT} = 4.272$ for the Intercooled Regenerative (ICR) gas turbine with reheat in Section 5.2. The other system-parameter values roughly are typical for regenerative-gas-turbine applications.

For the system-parameter values above, the criteria given by Equations 6.116 and 6.117 are

$$\alpha_R \tau_N / H^2 = \alpha_R \tau_X / H^2 = 34 \gg 1$$

Therefore, we conclude that a constant heat-transfer-coefficient ratio $[(h_{ef}/h)_{ss}]$ value can be used for some high-effectiveness parallel-plate regenerators to calculate regenerator effectiveness[9].

If the criteria given above are not satisfied, then an iterative approach must be used to calculate effectiveness. First, the fluid-flow temperatures must be calculated. Then, $(h_{ef}/h)(Fo_H)$ must be calculated for each axial location. New fluid-flow temperatures must then be calculated, considering these reductions in core heat-transfer performance. This cycle must be repeated until sufficient convergence has been achieved. An assumption in this method is that the

[9] For fully-developed laminar flows, the heat-transfer coefficient (h) is constant along the length of a passage. Therefore, $(h_{ef}/h)_{ss}$ is constant along the length of a passage for fully-developed laminar flows.

reductions in heat-transfer performance are small enough that the fluid-flow temperature profiles remain fully developed.

Because a steady-state value for the heat-transfer-coefficient ratio $[(h_{ef}/h)_{ss}]$ value can be used to calculate regenerator effectiveness for some high-effectiveness parallel-plate regenerators, a plot of $(h_{ef}/h)_{ss}$ vs. N_p is useful to the regenerator designer. Figure 6.17 shows such a plot.

Summary of Porosity Effects. The following conclusions can be drawn from the results given above.

1. Heat-transfer performance in a regenerator decreases with decreasing core porosity.

2. Heat-transfer performance can be correlated as a function of porosity number (N_p) via a heat-transfer-coefficient ratio (h_{ef}/h).

3. For many regenerator applications, a constant steady-state heat-transfer-coefficient ratio $[(h_{ef}/h)_{ss}]$ can be used, which leads to analysis simplification.

4. h_{ef}/h can be used to predict the heat-transfer-performance penalty incurred through low core porosity in a given application, or

5. h_{ef}/h can be used to calculate the increase in regenerator size required to achieve a given effectiveness specification.

6. Significant (although small) reductions in heat-transfer performance are incurred with cores of low conductivity (e.g. ceramic cores) and typical porosities ($p \approx 0.7$–0.8).

7. Reductions in heat-transfer performance due to core porosity for highly conductive cores (e.g. stainless-steel cores) are negigible for all but very low porosities ($p < 0.02$).

This section has shown that low core porosity can lead to a reduced regenerator heat-transfer performance. However, there are many incentives to reducing porosity, including lower carry-over leakage and a slower core-rotation rate. In addition, this section has shown that a high porosity and/or a high thermal conductivity are desirable for the core so that heat can easily be added to and extracted from the central portions of the passage walls. However, high core conduction in the other direction (in the direction of fluid flows) reduces regenerator effectiveness. Clearly an optimal porosity must result from a trade-off of all of these effects (and probably more, including manufacturing concerns). The current trend is toward high porosities ($p \approx 0.7$–0.8). For these cores, reductions in heat-transfer performance due to low porosity are small.

The regenerator designer who is considering specifying a low porosity ($p < 0.7$) for a ceramic core should consider the effects on heat-transfer performance described in this section.

6.1.11 Transient Operation

This section considers the transient responses of regenerator cores to cold start-ups. The key result is that the effectiveness of a regenerator responds in a dimensionless time period:

$$\tau^{\star}_{Xss} \equiv \frac{C_X \tau_{ss}}{(mc)_R} \sim 1 \; , \tag{6.118}$$

where $\tau^{\star}_{Xss}$ is the dimensionless time period; C_X is the greater of the two flow heat-capacity rates; τ_{ss} is the (dimensional) response time; and $(mc)_R$ is the heat capacity of the core. Equation 6.118 can be used to determine the time period, τ_{ss} required for the effectiveness of a regenerator to reach a steady-state value.

This section follows the work of Beck [5, 6]. First we give the theory. Then we show how Beck [5, 6] proved the theory through experiments with regenerator-core specimens.

Theory. The following gives the theory for the transient response of regenerator cores. First we consider single flow exposures, which are used to quantify the heat-transfer performance of regenerator cores. Then we consider effectiveness responses.

Single Flow Exposures. This section derives the theory for regenerator cores subjected to single flow exposures. With a single flow exposure, a regenerator core initially has a uniform temperature distribution, and is then instantaneously exposed to a flow with a different temperature. In addition to the simplifying assumptions given in Section 6.1.3, we assume no conduction in the core material. We assume this because regenerator-response time is nearly independent of axial core conduction for gas-turbine regenerators.

Under the given assumptions, one of the two governing equations for a single flow exposure is

$$\frac{\partial T}{\partial t^\star} + \frac{C}{(hA)}\frac{\partial}{\partial \zeta}\left(\frac{\partial T}{\partial t^\star}\right) + \frac{\partial T}{\partial \zeta} = 0 \; , \tag{6.119}$$

where T is the fluid temperature; $t^\star \equiv Ct/(mc)_R$, a dimensionless time; C is the flow heat-capacity rate; (hA) is the convective conductance; $\zeta \equiv x/L$ is a dimensionless coordinate in the flow direction; and L is the core thickness. Equation 6.119 applies for the fluid temperature. The second governing equation applies for the core material, but it is identical to Equation 6.119. The boundary conditions associated with these equations are

1. $T(\zeta; t^\star = 0)$, the initial temperature distribution in the fluid;
2. $T_R(\zeta; t^\star = 0)$, the initial temperature distribution in the core material; and
3. $T(\zeta = 0; t^\star)$, the inlet temperature of the flow.

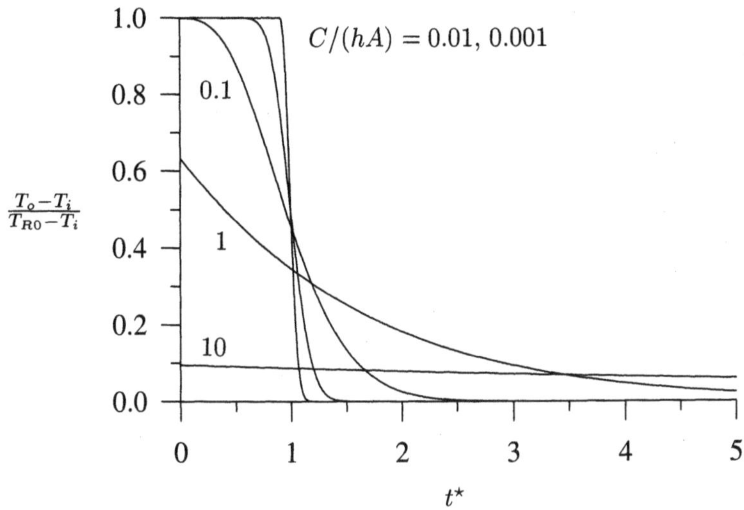

Figure 6.32: Responses of the Outlet Temperature of the Fluid Flow to Single Flow Exposures of a Regenerator-Core Passsage. Reprinted from Beck [5] with permission from the ASME.

Beck [5, 4] did not attempt an analytical solution to Equation 6.119, except for the limiting case as the ratio of the flow heat-capacity rate to the convective conductance approaches infinity: $C/(hA) \to \infty$ (which could be effected with a thin core). For this case, the core material acts as a lumped body, and the outlet-temperature response is

$$\frac{T_o - T_i}{T_{R0} - T_i} = \frac{(hA)}{C} \exp\left\{\frac{-(hA)t^\star}{C}\right\} , \tag{6.120}$$

which is a first-order response with a dimensionless time constant of $C/(hA)$.

For general $C/(hA)$ values, Beck [5, 6] solved Equation 6.119 numerically. Figure 6.32 shows plots of the resulting solutions. The integrals of the responses from $t^\star = 0$ to $t^\star = \infty$ are equal to one because the amount of heat transferred to the core is equal to the time integral of the rate at which the flow gives up its heat:

$$(T_{R0} - T_i)(mc)_R = C \int_0^\infty (T_o - T_i)dt . \tag{6.121}$$

For high $C/(hA)$ values, the plots in Figure 6.32 approach first-order responses given by Equation 6.120: for a fixed value of the flow heat-capacity rate (C), the convective conductance is small, so heat is removed slowly from (or added slowly to) the passage-tube walls. For low $C/(hA)$ values, the plots approach a step function:

$$\frac{T_o - T_i}{T_{R0} - T_i} = \begin{cases} 1 & \text{if } t^\star < 1 \\ 0 & \text{if } t^\star > 1 \end{cases} . \tag{6.122}$$

For low $C/(hA)$ values, heat is removed from (or added to) the passage-tube walls at the maximum rate (which is limited by the difference between the initial core temperature and the flow temperature, $T_{R0} - T_i$), until the core achieves the fluid flow's inlet temperature.

The designer can use the temperature responses to single flow exposures to determine the heat-transfer performance of regenerator cores. The designer is usually faced with one of two situations: (1) the heat capacity of a core specimen $[(mc)_R]$ is known, but the convective conductance $[(hA)]$ is unknown; or (2) the convective conductance is known from theory or otherwise, but the heat capacity of the core is unknown. In either of these situations, the designer should first determine the heat-capacity rate of the flow to which she or he exposed the core. Then, either the heat capacity of the core or the convective conductance should be determined. The unknown quantity (either the heat capacity or the convective conductance) should then be determined by matching the experimental results with the theory. Below we show how Beck [5, 6] tested ceramic regenerator cores using this procedure.

Effectiveness Responses. This section gives the theory for the transient response of regenerator effectiveness. We consider cold start-ups in which the core initially has a uniform temperature distribution and is then rotated between cold and hot flows in its manner of normal operation. We are interested in determining how long it will take for the effectiveness of the regenerator to reach its steady-state value. Section 6.1.4 gives appropriate governing equations and boundary conditions. Beck [5, 6] solved the equations numerically subject to the boundary conditions.

This section addresses effectiveness-response times (not effectiveness magnitudes). Therefore, we give results in terms of normalized effectivenesses, ϵ/ϵ_{ss}[10].

When we normalize time, normalized effectiveness responses are independent of the dimensionless core-rotation rate (C_{ROT}), except for low values of the product of the heat-capacity-rate ratio and the dimensionless core-rotation rate (e.g. $C_{RAT}C_{ROT} < 1$) for which the core rotates few times before steady-state effectiveness is reached. Normalized effectiveness responses are nearly independent of the dimensionless core size (NTU) and the heat-capacity-rate ratio. Also, normalized effectiveness responses are independent of the convective-conductance ratio ($(hA)'$) and of axial conduction (which depends on the parameters λ, SC, and As'). Figure 6.33 shows plots of normalized effectiveness responses for various NTU, and Figure 6.34 shows plots of responses for various C_{RAT}[11].

From the plots of normalized effectiveness responses, we infer the following criterion for the dimensionless time period to steady-state effectiveness:

$$\tau^{\star}_{X_{ss}} \equiv \frac{C_X \tau_{ss}}{(mc)_R} \sim 1 \; , \tag{6.123}$$

[10] Steady-state effectiveness results are identical to results of calculations performed using the method reported by Beck [7], which is given in Section 6.1.8.

[11] The steps in the plots result from simplification in the numerical-integration algorithm.

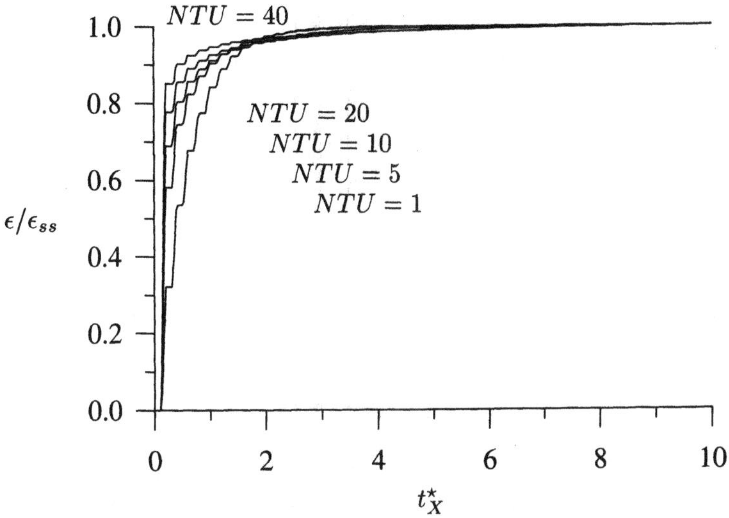

Figure 6.33: Normalized Effectiveness Responses for $C_{RAT} = (hA)' = 1$; $NTU = 1, 5, 10, 20$, and 40. Reprinted from Beck [5] with permission from the ASME.

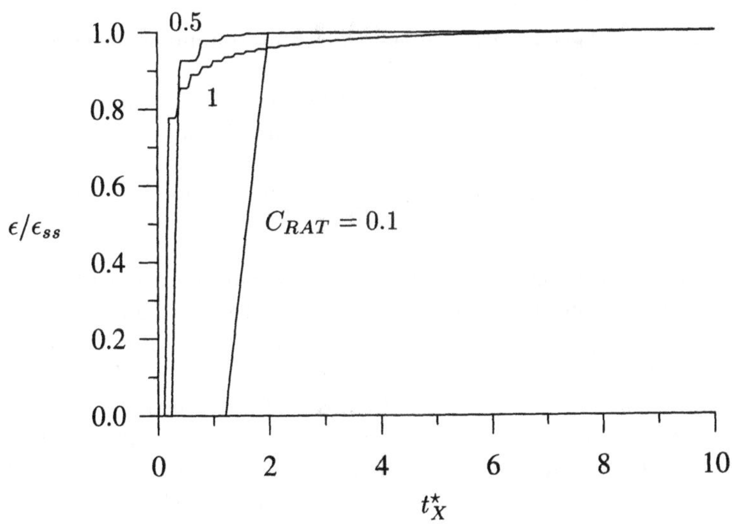

Figure 6.34: Normalized Effectiveness Responses for $(hA)' = 1$; $NTU = 20$; $C_{RAT} = 0.1, 0.5$, and 1. Reprinted from Beck [5] with permission from the ASME.

where C_X is the greater of the two heat-capacity rates (typically the heat-capacity rate of the exhaust in regenerative gas turbines); τ_{ss} is the (dimensional) time period to steady-state effectiveness; and $(mc)_R$ is the heat capacity of the core. In terms of dimensional time, this criterion can be expressed as

$$\tau_{ss} \sim \left\{ \frac{NTU[1+(hA)'][1+As']C_{RAT}}{4As'(1-SC)k_N} \right\} \times \left\{ \frac{(\rho c)_R D_H^2}{Nu_H} \left(\frac{1-p}{p} \right) \right\}, \tag{6.124}$$

where

NTU	=	Number of Transfer Units, a dimensionless core size (–);
$(hA)'$	=	Convective-conductance ratio (–), the ratio of the convective conductance on the compressed-air side (C_N side) to that on the exhaust side (C_X side);
As'	=	Solid-area ratio (–), the ratio of the solid area on the compressed-air side to that on the exhaust side;
C_{RAT}	=	Heat-capacity-rate ratio (–), the ratio of the heat-capacity rate on the compressed-air side to that on the exhaust side;
SC	=	Seal Coverage (–), the fraction of the core face area covered by seals;
k_N	=	Thermal conductivity of the compressed air (W/m-K);
$(\rho c)_R$	=	Specific heat capacity per unit volume of the core material ($J/m^3 - K$);
D_H	=	Hydraulic diameter of the core passages (m);
Nu_H	=	Nusselt number associated with the passage geometry (–); and
p	=	Core porosity (–), voids volume divided by total volume.

The regenerator designer determines values for the parameters NTU, $(hA)'$, As', C_{RAT} and SC in the first term of Equation 6.124 in different ways, depending on the design procedure that she or he uses. In the method of Kays and London [37] and in the method for direct design (see Section 5.1), NTU is calculated from a specified effectiveness, $(hA)'$ or As' are specified (one results from specification of the other), and C_{RAT} is calculated. In optimal design, NTU, $(hA)'$, As', C_{RAT}, and SC are determined through optimization of the regenerative gas turbine for steady-state operation (see Beck [4, 6] and Section 5.2). The principal contributor to the first term in Equation 6.124 is NTU: high-effectiveness regenerators with high values of NTU produce long response times. The parameters in the second term of Equation 6.124 are characteristics of the core material and of the core-passage geometries. For a fast effectiveness response, a highly porous core with small passages and a low specific heat capacity per unit volume should be used. The Nusselt number, which is a function of passage geometry, should also be large. For many practical cases (especially for ceramic cores), passage size and geometry and core porosity are determined principally by manufacturing considerations; however, Equation 6.124 should be considered when choosing a regenerator core for a particular application.

Example Calculation. For an example of the use of Equation 6.124, consider Wilson's [73] regenerator design. The relevant design data are:

$$
\begin{aligned}
NTU &= 22; \\
(hA)' &= 1; \\
As' &= 1.05; \\
C_{RAT} &= 0.990; \\
SC &= 0.2; \\
k_N &= 0.0569 \text{ W/m-K}; \\
(\rho c)_R &= 2.60 \text{ MJ/m}^3\text{-K}; \\
D_H &= 0.511 \text{ mm}; \\
Nu_H &= 2.65; \text{ and} \\
p &= 0.708.
\end{aligned}
$$

From Equation 6.124, the response time is on the order of 49 seconds: $\tau_{ss} \sim 49$. From Figure 6.33 we conclude that the effectiveness of Wilson's [73] regenerator would reach 90% of its steady-state value in about one minute ($t^\star = 60/49$) and practically 100% of its steady-state value in under eight minutes ($t^\star = 8{\times}60/49$).

Experimental Results. This section describes experiments performed by Beck [5, 6] that prove the theory for transient operation detailed in the previous sections. First we describe the apparatus that Beck [5, 6] used. With this apparatus, Beck [5, 6] proved the theory for transient operation; however, the apparatus is also generally useful for measuring the heat-transfer performance of regenerator cores.

After we describe the apparatus, we compare with the theory experimental results for single flow exposures and for effectiveness responses. Finally, we conclude that regenerator cores respond in a dimensionless time period, $\tau^\star_{Xss} \sim 1$.

Experimental Apparatus. Figure 6.35 shows a schematic of the experimental apparatus that Beck [5, 6] used to test regenerator-core specimens. Table 6.1 lists the components for the reader that would like to build a similar apparatus for testing regenerator cores. With the apparatus, single flow exposures as well as simulations of regenerator operation can be performed. For a single flow exposure, the central block of the test section (Component S in Table 6.1) is first moved to its left position and the specimen (G) is heated to a uniform temperature. Then the central block is quickly moved to its right position (the position shown in the schematic) and the core is cooled. For a simulation of a

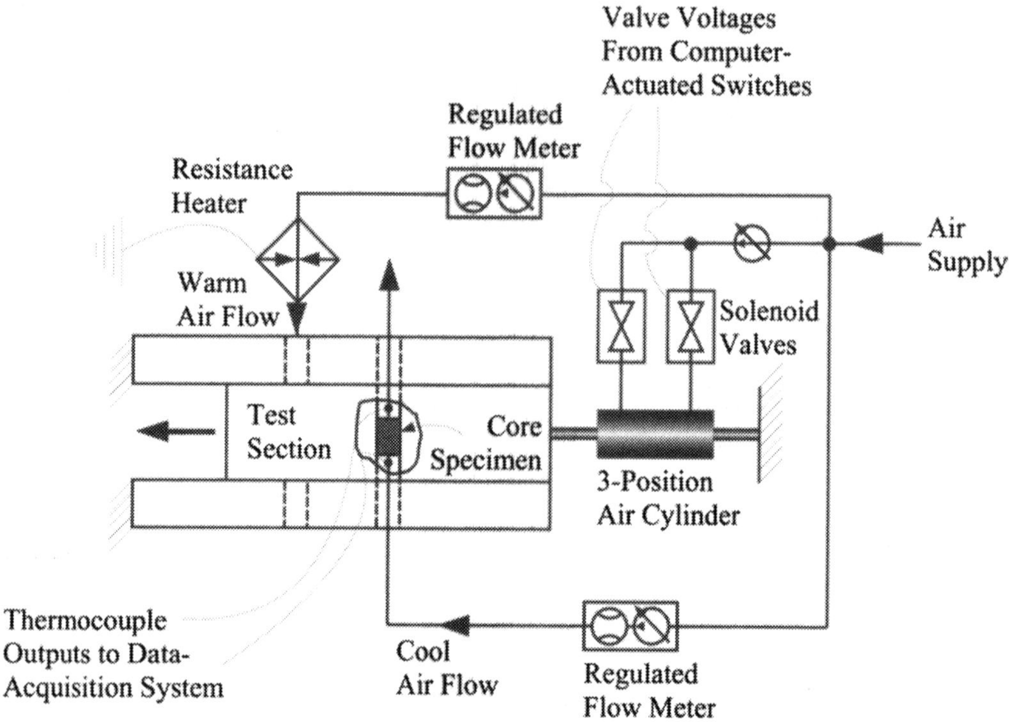

Figure 6.35: Schematic of an Experimental Apparatus for Tests of Heat-Transfer Performance. Reprinted from Beck [5] with permission from the ASME.

regenerator's operation, the core is alternately exposed to the cool and warm flows. Between exposures, the central block is moved to a central position in which both ends of the specimen are sealed. This simulates travel under seals in an actual regenerator.

Consider the cool-air flow. Air flows from the air supply, through a regulated flow meter (H), then through the core specimen. Thermocouples (L) at each end of the specimen are used to record the air temperature at the inlet and at the outlet of the specimen. The flow path is similar for the warm-air flow except the flow passes through an electrical-resistance heater (N) between a flow meter (C) and the specimen.

Consider the linear-actuator system. Commands are sent at specified intervals from software to move the central block of the test section. The commands travel via the input/output (I/O) board (Q) to the output modules (A). These output modules open and close circuits that include two-position solenoid valves (D). The valves direct the regulated (E) compressed-air flow to four ports on the linear-actuator cylinder (F). Thus, the linear actuator can be moved to any of three positions.

Consider the temperature-data acquisition system. The thermocouple mod-

Letter	Component Description
A	MetraByte SM-OAC5, 120 VAC output module
B	MetraByte SRA-01, solid-state I/O board
C	Dwyer 100 SCFH flow meter
D	Humphrey H180-4E1, two-position five-port solenoid valves
E	Watts R384-02 C, miniature regulator
F	Allenair APD $1\frac{1}{2} \times 3 \times 3$, pneumatic-linear-actuator cylinder
G	Corning cordierite cylindrical regenerator-core specimen
H	Dwyer 50 SCFH flow meter
I	MetraByte PWR-51A, 5-V 1-AMP power supply
J	MetraByte MB-01, direct-connection module mounting rack
K	MetraByte MB37-T, T-type thermocouple module
L	Omega 0.003-inch T-type thermocouple wire
M	MetraByte C-8MB1, DAS-8 to MB-01 rack cable
N	Electrical-resistance heater
O	MetraByte DAS-8, eight-channel A/D interface
P	NPC 486 VLB/66 computer
Q	MetraByte PIO-12, 24-line parallel I/O board
R	MetraByte C1800, 37-conductor 18-inch cable
S	Test section

Table 6.1: Descriptions of Components in the Experimental Apparatus. Reprinted from Beck [5] with permission from the ASME.

Specimen Number	L (mm)	N_{pass}	(hA) (W/K)	$(mc)_R$ (J/K)
1	67	68	1.91	14.1
2	67	405	14.6	19.9
3	66	145	4.49	17.1
4	60	209	5.25	8.93

Table 6.2: Core-Specimen Data. Reprinted from Beck [5] with permission from the ASME.

ules (K) process the thermocouple voltages. The resulting voltages travel to the analog-to-digital (A/D) interface (O), and the software records them.

Figure 6.36 shows end-view enhanced[12] pictures of the core specimens that Beck [5, 6] tested. The convective conductances were calculated from the conductivity of air ($k_f \approx 0.029$ W/m-K) and from data for the thicknesses of the specimens (L) and for the number of passages that each specimen contained. Table 6.2 lists these data. For square passages (Specimens 1 and 4),

$$(hA) = 4k_f L N_{pass} Nu \; ; \tag{6.125}$$

for 2 × 1 rectangle passages (Specimen 2),

$$(hA) = 4.5k_f L N_{pass} Nu \; ; \tag{6.126}$$

and for equilateral-triangle passages (Specimen 3),

$$(hA) = 5.20k_f L N_{pass} Nu \; . \tag{6.127}$$

Table 6.2 lists the resulting core-specimen data.

The next section details how Beck [5, 6] measured the heat capacities of the core specimens (which are listed in Table 6.2) with single flow exposures.

Single Flow Exposures. Beck [5, 6] used single flow exposures to determine the heat capacities of all of the core specimens. For all runs, the specimens were exposed to a warm air flow of $\dot{V}_{warm} = 7.87E-4$ m^3/s then a cool air flow of $\dot{V}_{cool} = 3.93E-4$ m^3/s. Figure 6.37 shows a typical response and the associated theoretical response. The experimental responses are slower for high values of dimensionless time [$t^\star \equiv Ct/(mc)_R$] due to conduction from the specimen peripheries. Also, the convective heat-transfer coefficient (h) has a lower value for a constant-wall-temperature boundary condition than for a boundary condition of a constant wall heat flux. For small time, the temperature distribution in the core material is uniformly equal to the initial temperature distribution, T_{R0}. For large time, the temperature profile in the core material is uniformly the flow inlet temperature, T_i. Therefore, the experimental responses

[12] Beck [5, 6] enhanced the pictures to delineate between passages and passage tubes.

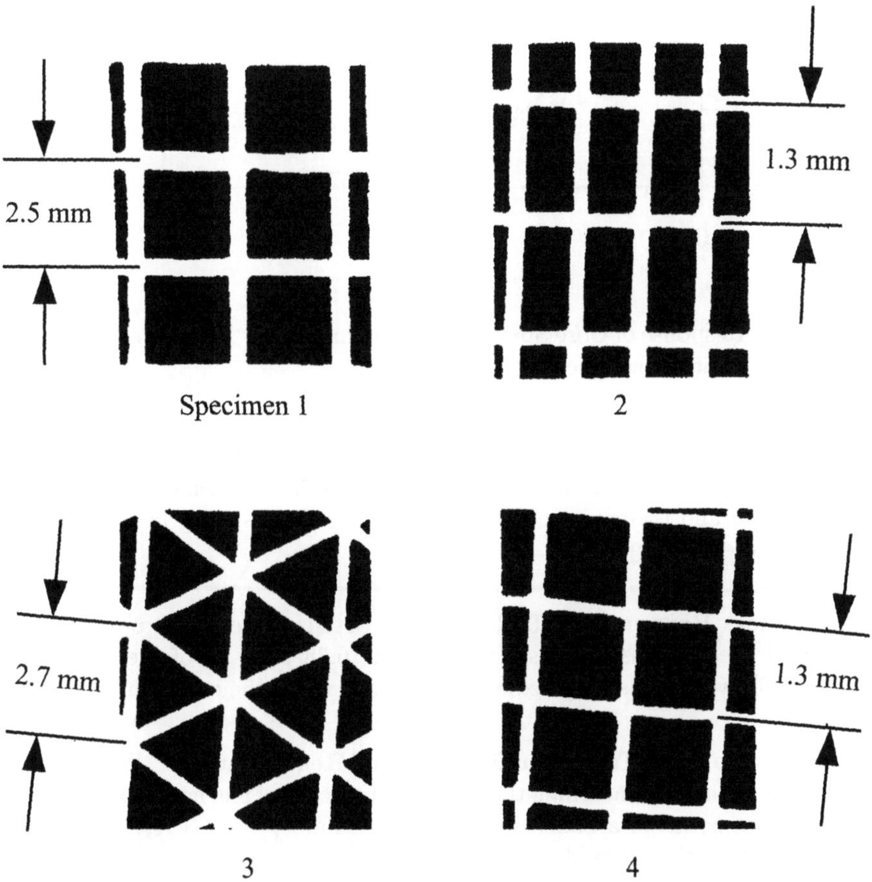

Figure 6.36: End-View Enhanced Pictures of Magnified Core Specimens. Reprinted from Beck [5] with permission from the ASME.

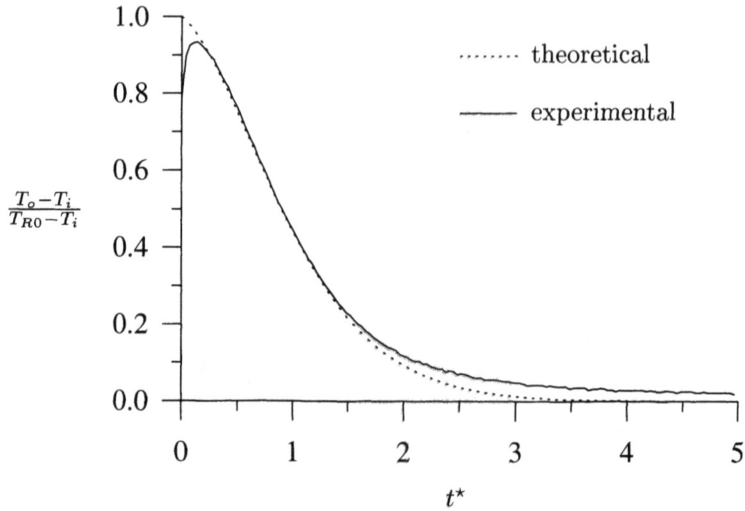

Figure 6.37: Response of Core Specimen 1 to a Single Flow Exposure. Reprinted from Beck [5] with permission from the ASME.

are slower than the theoretical responses for small and large times due to the reduced convective heat-transfer coefficient. Otherwise, agreement between experimentation and theory is good. Table 6.2 lists the resulting heat capacities of the core specimens.

Effectiveness Responses. Table 6.3 gives details of Beck's [5, 6] twelve experimental effectiveness-response runs. Figure 6.38 plots all of the data from Beck's [5, 6] twelve runs. The experimental data verify the criterion $\tau^\star_{X ss} \sim 1$.

Conclusions on the Transient Response of Regenerators. We draw the following conclusions concerning the transient response of regenerators. A regenerator core responds in a dimensionless time period $\tau^\star_{X ss} \equiv C_X \tau_{ss}/(mc)_R \sim 1$, or a dimensional time period

$$\tau_{ss} \sim \left\{ \frac{NTU[1 + (hA)'][1 + As']C_{RAT}}{4As'(1 - SC)k_N} \right\} \times \left\{ \frac{(\rho c)_R D_H^2}{Nu_H} \left(\frac{1 - p}{p} \right) \right\} \quad (6.128)$$

This response time is a function of parameters that are fixed for a given regenerator design (see the discussion below Equation 6.124 for details). It is independent of core-rotation rate, which is the only practically available control input. Therefore, the response time is a function of only the flows, the core size, and other physical parameters, and it cannot be reduced with a control system.

Run Number	Specimen Number	$\dot{V}_N$ (m^3/s)	$\dot{V}_X$ (m^3/s)	τ_N (s)	$\tau_S/2$ (s)	τ_X (s)
1	1	3.93E-4	7.87E-4	5	1	5
2	1	3.93E-4	7.87E-4	10	1	5
3	1	3.93E-4	7.87E-4	15	1	5
4	2	3.93E-4	7.87E-4	5	1	5
5	2	3.93E-4	7.87E-4	10	1	5
6	2	3.93E-4	7.87E-4	15	1	5
7	3	3.93E-4	7.87E-4	5	1	5
8	3	3.93E-4	7.87E-4	10	1	5
9	3	3.93E-4	7.87E-4	15	1	5
10	4	3.93E-4	7.87E-4	5	1	5
11	4	3.93E-4	7.87E-4	10	1	5
12	4	3.93E-4	7.87E-4	15	1	5

Table 6.3: Details of Experimental Effectiveness-Response Runs. Reprinted from Beck [5] with permission from the ASME.

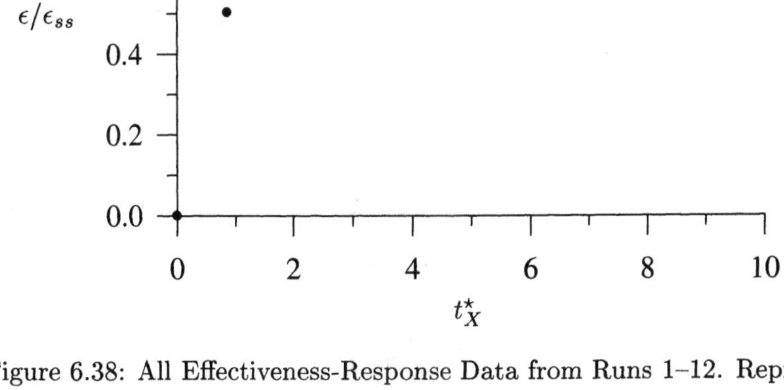

Figure 6.38: All Effectiveness-Response Data from Runs 1–12. Reprinted from Beck [5] with permission from the ASME.

6.1.12 Seal Shape

These last two sections on heat transfer describe effects on regenerator effectiveness of two sources of flow non-uniformity. In general, flow non-uniformity can have four sources:

1. Non-ideal headers;
2. Non-uniform passages;
3. Seals of arbitrary shape; and
4. Discontinuous core rotation.

Each source affects the heat-transfer performance and pressure-drop performance of a regenerator in different ways. These last two sections address only the last two sources of flow non-uniformity (seals of arbitrary shape and discontinuous core rotation) because headers that are nearly ideal can be designed and current manufacturing techniques produce nearly uniform core passages. However, the methods given in these last two sections indicate how the effects of flow non-uniformity from any of the four sources can be quantified.

This section addresses the effects on heat-transfer performance of flow non-uniformity due to seal shape in regenerators with axial flows (see Chapter 4 for descriptions of axial-flow and radial-flow regenerators). The principal conclusion is that effectiveness can be reduced by a few percentage points due to the shape of the face seals that separate the flows (and partially cover the core), under the following *unfavorable* conditions:

1. Highly unequal flow areas;
2. Wide seals;
3. A small ratio of the inner radius of the core to the outer radius;
4. High-effectiveness regenerators with large values for the Number of Transfer Units (NTU); and
5. Equal (or nearly equal) flow heat-capacity rates (a heat-capacity-rate ratio of $C_{RAT} \approx 1$).

The rest of this section gives analyses that quantify the effects that these unfavorable conditions have on regenerator effectiveness. In our analyses we assume infinite core-rotation rate, zero axial conduction, and steady-state operation.

Method. In the following, we compare the effectivenesses of rotary regenerators with uniform-width seals with otherwise-identical (with identical values for the overall heat-transfer parameters) regenerators with sector-shape seals. Consider the regenerator core represented in Figure 6.39. The sector seals are sized so that the compressed-air and exhaust sides have the same areas as for

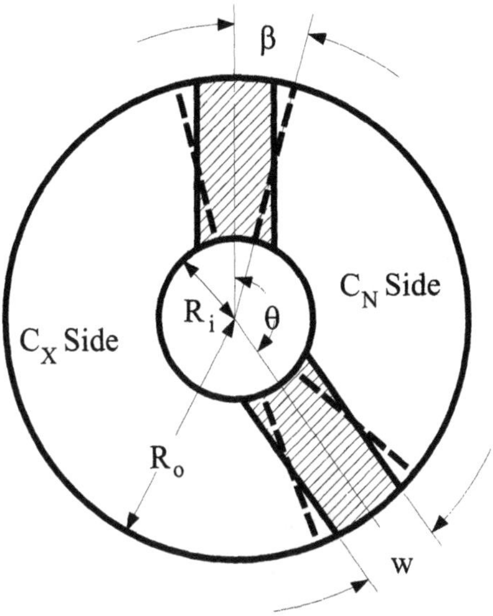

Figure 6.39: Axial-View Schematic of an Axial-Flow Rotary-Regenerator Core with Either Uniform-Width Seals (Solid Lines and Crosshatched area) or Sector Seals (Dashed Lines).

the uniform-width-seals case. The only differences in the uniform-width-seal regenerator and the one with sector seals are due to seal shape.

To match the area covered by a sector-shaped seal to that of a uniform-width seal, the parameter β, defined in Figure 6.39, is chosen appropriately. The area of half of one of the sector seals can be expressed as follows.

$$\frac{A_{sect}}{2} = \left(\frac{R_o^2 - R_i^2}{2}\right)\beta \tag{6.129}$$

The area of half of one of the uniform-width seals can be expressed as

$$\frac{A_W}{2} = \left(\frac{w}{2}\right)(R_o - R_i) + \Delta_2 - \Delta_1 \tag{6.130}$$

where Δ_1 and Δ_2 are defined in Figure 6.40. To calculate Δ_1 and Δ_2 the angle $\phi(r)$ (shown in Figure 6.41) is first defined. This angle can be calculated from

$$\phi\left(r\right) = \csc^{-1}\left(2\left[\frac{r}{w}\right]\right) \tag{6.131}$$

Therefore, $\phi(r)$ can be referred to as $\phi(r/w)$.

Now consider the areas represented in Figure 6.42. These areas are associated with those in Figure 6.40, as indicated by the horizontal and vertical cross-hatching. As shown in Figure 6.42, Δ_1 and Δ_2 can be expressed as

$$\Delta_1 = \frac{wR_o}{2} - \frac{wR_o cos\left(\phi\left(\frac{R_o}{w}\right)\right)}{4} - \frac{R_o^2\phi\left(\frac{R_o}{w}\right)}{2} \tag{6.132}$$

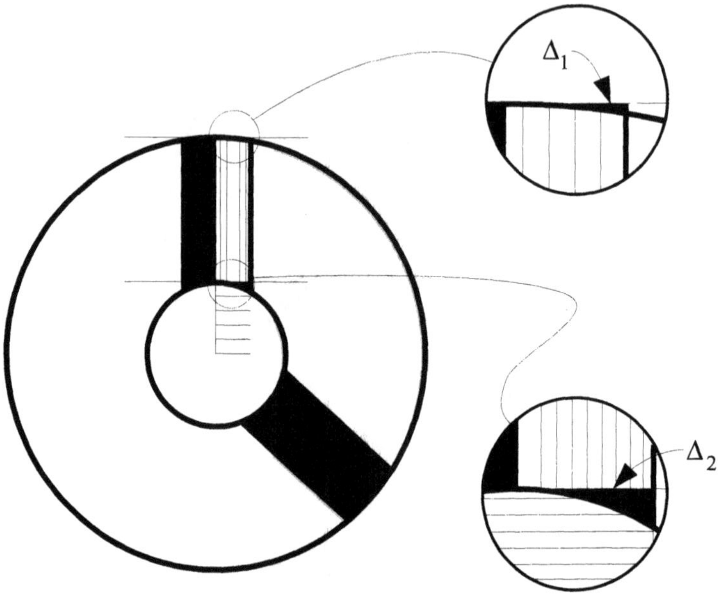

Figure 6.40: Definitions of Δ_1 and Δ_2.

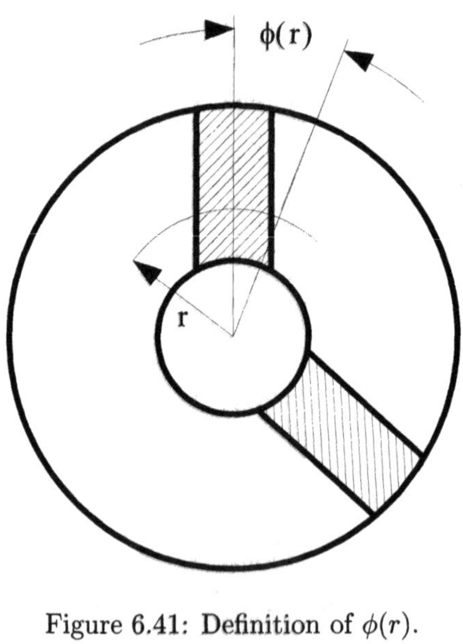

Figure 6.41: Definition of $\phi(r)$.

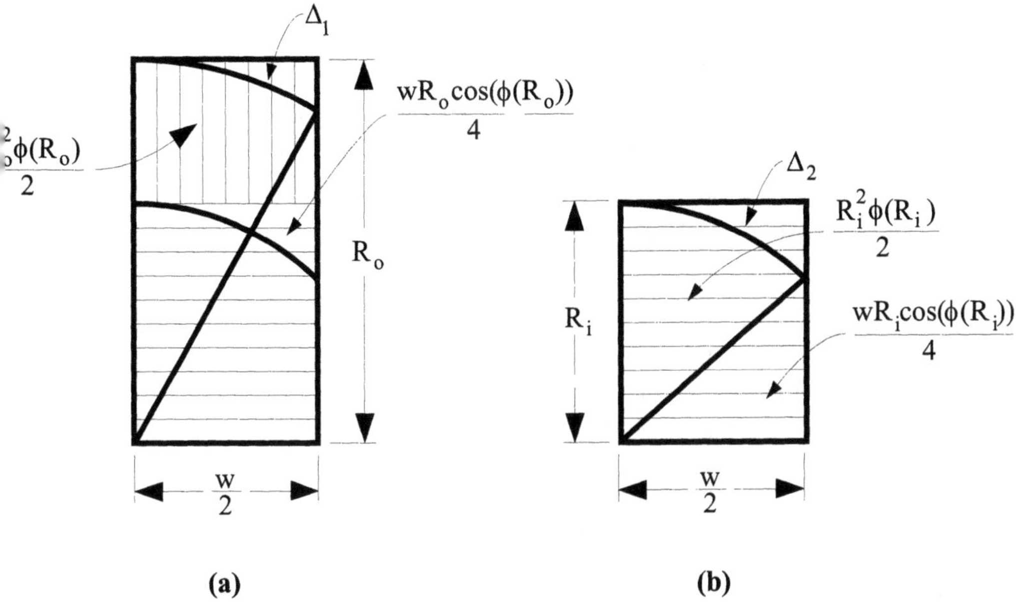

Figure 6.42: (a) Relevant Areas Used to Calculate Δ_1 and (b) Relevant Areas Used to Calculate Δ_2.

and

$$\Delta_2 = \frac{wR_i}{2} - \frac{wR_i cos\left(\phi\left(\frac{R_i}{w}\right)\right)}{4} - \frac{R_i^2\phi\left(\frac{R_i}{w}\right)}{2} \tag{6.133}$$

or

$$\Delta_1 = \left(\frac{R_o^2}{2}\right)\left(\left[\frac{w}{R_o}\right]\left[1 - \frac{cos\left(\phi\left(\frac{R_o}{w}\right)\right)}{2}\right] - \phi\left(\frac{R_o}{w}\right)\right) \tag{6.134}$$

and

$$\begin{aligned}\Delta_2 \quad &= \quad \left(\frac{R_i^2}{2}\right)\left(\left[\frac{w}{R_o}\right]\left[\frac{R_o}{R_i}\right]\left[1 - \frac{cos\left(\phi\left(\left[\frac{R_i}{R_o}\right]\left[\frac{R_o}{w}\right]\right)\right)}{2}\right] - \right.\\ &\left.\phi\left(\left[\frac{R_i}{R_o}\right]\left[\frac{R_o}{w}\right]\right)\right)\end{aligned} \tag{6.135}$$

in terms of the dimensionless parameters, w/R_o and R_i/R_o.

Setting $A_{sect} = A_w$ gives the following expression for β.

$$\begin{aligned}\beta \quad &= \quad \left(\frac{w}{R_o}\right)\left(\frac{1}{1+\frac{R_i}{R_o}}\right) + \\ &\left(\frac{1}{1-\left[\frac{R_i}{R_o}\right]^2}\right)\left(\left[\frac{R_i}{R_o}\right]^2\left[\left\{\frac{w}{R_o}\right\}\left\{\frac{R_o}{R_i}\right\}\right.\right.\end{aligned}$$

$$\left\{1-\frac{\cos\left(\phi\left(\left\{\frac{R_i}{R_o}\right\}\left\{\frac{R_o}{R_i}\right\}\right)\right)}{2}\right\}-\phi\left(\left(\frac{R_i}{R_o}\right)\left(\frac{R_o}{w}\right)\right)\Bigg]$$
$$-\left[\left\{\frac{w}{R_o}\right\}\left\{1-\frac{\cos\left(\phi\left(\frac{R_o}{w}\right)\right)}{2}\right\}-\phi\left(\frac{R_o}{w}\right)\right]\Bigg) \tag{6.136}$$

Θ Derivation. To calculate the relevant heat-transfer parameters for uniform-width-seal regenerators as functions of radial location, expressions for the differential areas on each side of a regenerator's seals as functions of radial location are required. A parameter that facilitates derivation of these expressions is θ, which is defined in Figure 6.39. (This parameter results from the designer's choice of overall conductance ratio).

For a regenerator with sector seals, the solid cross-sectional area on the C_N side can be expressed as

$$As_N = \pi\left(R_o^2 - R_i^2\right)\left(\frac{\theta - 2\beta}{2\pi}\right)(1-p) \tag{6.137}$$

Similarly, the solid area on the C_X side can be expressed as

$$As_X = \pi\left(R_o^2 - R_i^2\right)\left(\frac{2\pi - \theta - 2\beta}{2\pi}\right)(1-p) \tag{6.138}$$

Then the solid-area ratio, As', for the sector-seal regenerator is

$$As' = \frac{\theta - 2\beta}{2\pi - \theta - 2\beta} \tag{6.139}$$

This equation can be solved for θ to yield

$$\theta = \frac{2\pi As' + 2\left(1 - As'\right)\beta}{\left(1 + As'\right)} \tag{6.140}$$

C_{RAT_W} Derivation. The heat-capacity-rate ratio, C_{RAT}, for a sector-seal regenerator is defined as

$$C_{RAT} \equiv \left(\frac{c_{p_N}}{c_{p_X}}\right)\left(\frac{\dot{m}_N}{\dot{m}_X}\right) \tag{6.141}$$

The fluid specific heats are assumed constant (see Section 6.1.3). Therefore, the ratio, c_{p_N}/c_{p_X} is independent of radial location in the uniform-width-seal case, and this ratio is the same for a uniform-width-seal regenerator and an otherwise-identical one with sector seals.

The entering-fluid mass fluxes are assumed uniform over the flow-inlet cross section. For a sector-seal regenerator, the ratio of the flow area on the C_N side to that on the C_X side is constant with radial location and equal to $(\theta - 2\beta)/(2\pi - \theta - 2\beta)$, the solid-area ratio. Therefore, the mass-flow-rate ratio, $\dot{m}_N/\dot{m}_X$, and C_{RAT} are constant with radial location. For a regenerator with uniform-width seals, however, this mass-flow-rate ratio varies with radial location. For this kind of regenerator, the differential mass flow rate on the C_{N_W} side at a radial location can be expressed as

$$\delta \dot{m}_{N_W}(r) = \dot{m}_N \left(\frac{\delta As_{N_W}(r)}{As_N} \right) \tag{6.142}$$

where $\delta As_{N_W}(r)$ is the differential solid area on the C_{N_W} side and is equal to

$$\delta As_{N_W}(r) = (1-p)r(\theta - 2\phi)\delta r \tag{6.143}$$

Similarly,

$$\delta \dot{m}_{X_W}(r) = \dot{m}_X \left(\frac{\delta As_{X_W}(r)}{As_X} \right) \tag{6.144}$$

where

$$\delta As_{X_W}(r) = (1-p)r(2\pi - \theta - 2\phi)\delta r \tag{6.145}$$

Both As_N and As_X are identical for the uniform-width-seal and the sector-seal cases through the derivation of β, described earlier[13].

Using the above relations, the mass-flow-rate ratio for a uniform-width-seal regenerator is

$$\frac{\dot{m}_{N_W}}{\dot{m}_{X_W}} \left(\frac{r}{w} \right) = \left(\frac{\dot{m}_N}{\dot{m}_X} \right) \left(\frac{\theta - 2\phi}{\theta - 2\beta} \right) \left(\frac{2\pi - \theta - 2\beta}{2\pi - \theta - 2\phi} \right) \tag{6.146}$$

Define

$$C'_{RAT_W} \left(\frac{r}{w} \right) \equiv C_{RAT} \left(\frac{\theta - 2\phi}{\theta - 2\beta} \right) \left(\frac{2\pi - \theta - 2\beta}{2\pi - \theta - 2\phi} \right) \tag{6.147}$$

This parameter could have a value greater than 1; however, a regenerator's heat-capacity-rate ratio must be less than one by definition. Therefore, calculate C_{RAT_W} as follows.

$$C_{RAT_W} \left(\frac{r}{w} \right) = \begin{cases} C'_{RAT_W} & \text{if } C'_{RAT_W} \leq 1 \\ \frac{1}{C'_{RAT_W}} & \text{if } C'_{RAT_W} > 1 \end{cases} \tag{6.148}$$

Physically, if $C'_{RAT_W}(r/w) > 1$, then the uniform-width seal regenerator has a C_{N_W} side at radial location r/w that is the C_X side of the sector-seal regenerator. Also, the C_{X_W} side at that radial location is the C_N side of the sector-seal regenerator. In other words, a regenerator's sides of minimum and maximum heat-capacity rate can switch at some radial locations (for the local annular flow) if uniform-width seals are used instead of sector seals.

[13] It will be shown that these areas can switch at some radial locations: $\delta As_{N_W} \in As_X$ (δAs_{N_W} is a part of As_X) and $\delta As_{X_W} \in As_N$ (δAs_{X_W} is a part of As_N) at some radial locations. The subsequent expression for C_{RAT_W} allows for this switching and is identical to the solution obtained if these areas are switched at the outset.

$(hA)'_W$ Derivation. For a sector-seal regenerator, the convective-conductance ratio, $(hA)'$, can be expressed as follows.

$$(hA)' = \left(\frac{h_N}{h_X}\right) As' \tag{6.149}$$

The ratio h_N/h_X is constant with radial location. Therefore, the convective-conductance ratio for a uniform-width-seal regenerator is

$$(hA)'_W\left(\frac{r}{w}\right) = \begin{cases} \left(\frac{h_N}{h_X}\right)\left(\frac{\delta As_{N_W}}{\delta As_{X_W}}\right) & \text{if } C'_{RAT_W} \le 1 \\ \left(\frac{h_X}{h_N}\right)\left(\frac{\delta As_{X_W}}{\delta As_{N_W}}\right) & \text{if } C'_{RAT_W} > 1 \end{cases} \tag{6.150}$$

This expression allows for switching of the sides of minimum and maximum heat-capacity rate at some radial locations when uniform-width seals are used.

The expressions for δAs_{N_W} and δAs_{X_W} derived above can be substituted into Equation 6.150 to yield

$$(hA)'_W\left(\frac{r}{w}\right) = \begin{cases} \left(\frac{h_N}{h_X}\right)\left(\frac{\theta-2\phi}{2\pi-\theta-2\phi}\right) & \text{if } C'_{RAT_W} \le 1 \\ \left(\frac{h_X}{h_N}\right)\left(\frac{2\pi-\theta-2\phi}{\theta-2\phi}\right) & \text{if } C'_{RAT_W} > 1 \end{cases} \tag{6.151}$$

and, using Equation 6.149,

$$(hA)'_W\left(\frac{r}{w}\right) = \begin{cases} \left[\frac{(hA)'}{As'}\right]\left[\frac{\theta-2\phi}{2\pi-\theta-2\phi}\right] & \text{if } C'_{RAT_W} \le 1 \\ \left[\frac{As'}{(hA)'}\right]\left[\frac{2\pi-\theta-2\phi}{\theta-2\phi}\right] & \text{if } C'_{RAT_W} > 1 \end{cases} \tag{6.152}$$

***NTU* Derivation.** The heat-transfer parameter NTU, which quantifies nondimensionally a regenerator's size, can be expressed as follows for sector-seal regenerators.

$$NTU = \left[\frac{h_N A_N}{c_{p_N}\dot{m}_N}\right]\left[\frac{1}{1+(hA)'}\right] \tag{6.153}$$

For a uniform-width-seal regenerator, the parallel parameter can be expressed as

$$NTU_W\left(\frac{r}{w}\right) = \begin{cases} \left[\frac{h_N\delta A_{N_W}}{c_{p_N}\delta\dot{m}_{N_W}}\right]\left[\frac{1}{1+(hA)'_W}\right] & \text{if } C'_{RAT_W} \le 1 \\ \left[\frac{h_X\delta A_{X_W}}{c_{p_X}\delta\dot{m}_{X_W}}\right]\left[\frac{1}{1+(hA)'_w}\right] & \text{if } C'_{RAT_W} > 1 \end{cases} \tag{6.154}$$

and, using Equation 6.153,

$$NTU_W\left(\frac{r}{w}\right) = \begin{cases} NTU\left[\frac{1+(hA)'}{1+(hA)'_W}\right] & \text{if } C'_{RAT_W} \le 1 \\ NTU\left[\frac{C_{RAT}}{(hA)'}\right]\left[\frac{1+(hA)'}{1+(hA)'_W}\right] & \text{if } C'_{RAT_W} > 1 \end{cases} \tag{6.155}$$

C_{ROTW} **Derivation.** The dimensionless core-rotation rate, C_{ROT}, quantifies the regenerator core's rotational speed. For a sector-seal regenerator, this parameter can be expressed as

$$C_{ROT} = \frac{c_R \dot{m}_R}{c_{p_N} \dot{m}_N} \tag{6.156}$$

where c_R is the core material's specific heat and $\dot{m}_R$ is the mass flow rate of the core material. An expression for the mass flow rate of the core material is

$$\dot{m}_R = \rho_R (1-p) \omega L \overline{R} (R_o - R_i) \tag{6.157}$$

where ρ_R is the core-material density; ω is the angular rotational speed; L is the core thickness; and $\overline{R}$ is an average radius. For Equation 6.157 to hold,

$$(\rho_R \omega L) \overline{R} (R_o - R_i) = \int_{R_i}^{R_o} \rho_R \omega L r dr \tag{6.158}$$

or

$$\overline{R} = \frac{R_o + R_i}{2} \tag{6.159}$$

Thus

$$C_{ROT} = \frac{c_R (\rho_R \omega L)(1-p)(R_o^2 - R_i^2)}{2 c_{p_N} \dot{m}_N} \tag{6.160}$$

The parallel parameter for a uniform-width-seal regenerator can be expressed as

$$C_{ROTW}\left(\frac{r}{w}\right) = \begin{cases} \frac{c_R \delta \dot{m}_{RW}}{c_{p_N} \delta \dot{m}_{NW}} & \text{if } C'_{RATW} \leq 1 \\ \frac{c_R \delta \dot{m}_{RW}}{c_{p_X} \delta \dot{m}_{XW}} & \text{if } C'_{RATW} > 1 \end{cases} \tag{6.161}$$

The differential core-material mass-flow rate at a radial location is

$$\delta \dot{m}_{RW}\left(\frac{r}{w}\right) = (\rho_R \omega L)(1-p) r \delta r \tag{6.162}$$

The differential fluid mass flow rates as functions of radial location are given by Equations 6.142 and 6.144. Also, from Equation 6.160,

$$c_{p_N} \dot{m}_N = \frac{c_R (\rho_R \omega L)(1-p)(R_o^2 - R_i^2)}{2 C_{ROT}} \tag{6.163}$$

Substituting these expressions into Equation 6.161,

$$C_{ROTW}\left(\frac{r}{w}\right) = \begin{cases} C_{ROT}\left(\frac{\theta - 2\beta}{\theta - 2\phi}\right) & \text{if } C'_{RATW} \leq 1 \\ C_{ROT} C_{RAT}\left(\frac{2\pi - \theta - 2\beta}{2\pi - \theta - 2\phi}\right) & \text{if } C'_{RATW} > 1 \end{cases} \tag{6.164}$$

Effectiveness Calculation. The expressions derived for the parameters above can be used to calculate the effectiveness of uniform-width-seal regenerators. Again, a regenerator's sides of minimum and maximum heat-capacity rate can switch when uniform-width seals are used. Therefore, the following definition is used.

$$\epsilon'_W\left(\frac{r}{w}\right) \equiv \begin{cases} \epsilon''_W & \text{if } C'_{RAT_W} \leq 1 \\ C_{RAT_W}\epsilon''_W & \text{if } C'_{RAT_W} > 1 \end{cases} \tag{6.165}$$

where ϵ''_W is calculated from C_{RAT_W}, $(hA)'_W$, NTU_W, and C_{ROT_W}.

The effectiveness will vary with radial location as $\phi(r/w)$ varies. One average effectiveness is desired. Effectiveness can be calculated from the core-inlet fluid temperatures and from the average C_N-side core-outlet temperature. The mass fluxes are assumed constant over the core face areas. Therefore, an average-effectiveness can be obtained from

$$\epsilon_W = \int_0^1 \epsilon'(r/w)\, d\left(\frac{A_{N_W}}{A_N}\right) \tag{6.166}$$

For general C_{ROT} values, no known closed-form expression for effectiveness exists: it must be calculated numerically (see Lambertson [40]). For $C_{ROT} \to \infty$, a closed-form solution does exist. This solution is given by Equation 6.1.

To calculate effectiveness for uniform-width-seal regenerators with $C_{ROT_W} \to \infty$, a computer program was used to integrate Equation 6.166 numerically. In this program, r/w was incremented from $(R_i/R_o)(R_o/w)$ to R_o/w, while C_{RAT_W}, $(hA)'_W$, and NTU_W were calculated at each r/w. These values were used to compute $\epsilon'_W(\delta A_{N_W}/A_N)$ values, which were summed to approximate ϵ_W.

To calculate effectiveness for uniform-width-seal regenerators with finite C_{ROT}, the numerical integration was performed by hand. For this case, r/w was incremented from $(R_i/R_o)(R_o/w)$ to R_o/w, while C_{RAT_W}, $(hA)'_W$, NTU_W, and C_{ROT_W} were calculated at each r/w. These values were used to compute $\epsilon'_W(\delta A_{N_W}/A_N)$ values using a computer-program implementation of Lambertson's [40] method. The sum of these values approximated ϵ_W.

Results. The method described in the last subsection was used to calculate effectiveness for uniform-width-seal regenerators for a variety of input parameters. A C computer program was used to perform the calculations. A listing of this program is given in Reference [6]. For $C_{ROT} \to \infty$, the entire calculations were performed with the program. For finite C_{ROT}, parameters were calculated and returned for incremented radial locations so that effectivenesses could be obtained for these locations. These parameters were then used for calculation of uniform-width-seal effectiveness. Results of calculations for $C_{ROT} \to \infty$ and one calculation for finite C_{ROT} are given below.

First we give a description of the mode of presentation. Then we show effects of using uniform-width seals instead of sector seals for a variety of regenerator configurations. We compare effectivenesses of uniform-width-seal regenerators with those of sector-seal regenerators for various values of the following heat-transfer parameters: As', NTU, C_{RAT}, and $(hA)'$. In addition, we assume an infinite core-rotation rate ($C_{ROT} \to \infty$). Finally, we consider a finite core-rotation rate and give results of an example calculation.

Mode of Presentation. We express the principal results in terms of a uniform-width-seal effect, defined as

$$\Delta_W \equiv \frac{\epsilon_W - \epsilon}{\epsilon} , \qquad (6.167)$$

where ϵ_W is the effectiveness of a regenerator with uniform-width seals and ϵ is the effectiveness of an otherwise-identical, sector-seal regenerator. Thus, negative values for Δ_W are associated with reductions in effectiveness when uniform-width seals are used instead of sector seals.

The uniform-width-seal effect is similar to the axial-conduction effect used by Bahnke and Howard [2] (see Section 6.1.7). Results expressed in terms of such a parameter can be used by researchers or designers. First, the designer should assume sector seals will be used, and analyze or design a regenerator under this simplification. Then, the results of the method given in this section can be used to determine the uniform-width-seal effect for the configuration in question. If the uniform-width-seal effect is significant, then either NTU should be increased, or one of the other parameters should be adjusted to effect a higher effectiveness to compensate; or the design should be modified to avoid a significant uniform-width-seal effect.

The principal results are given as plots of Δ_W vs. w/R_o, indexed by R_i/R_o.

Significance of w/R_o and R_i/R_o. The parameter w/R_o is significant because differences in effectiveness occur due to differences in the amount of time passages at different radial locations spend exposed to each flow. Consider the two extreme-configuration cases represented in Figure 6.43[14]. For case *a*, the amount of time that each passage spends exposed to each flow differs little if uniform-width seals are used instead of sector seals. For case *b*, however, differences are great. Passages close to the hub (passages for which $r \approx R_i$) spend a vanishingly small amount of time on the C_N side. Therefore, these passages have effectivenesses that approaches unity. Passages close to the rim (passages for which $r \approx R_o$) spend a significantly larger portion of the time exposed to the C_N side than do the same passages if sector seals are used. Therefore, these passages have significantly lower effectivenesses if uniform-width seals are used. In general, when uniform-width seals are used, the effectivenesses of passages decrease with increases in the radius at which the passages are located. The number of passages increases with radius. Therefore, the lower-effectiveness outer passages have more weight in determining the uniform-width-seal regenerator's overall effectiveness. In general, uniform-width-seal regenerators have

[14] The solid-area ratio, As', is assumed less than unity for discussion of effectiveness variations with radial location in this section. All trends identified are reversed for $As' > 1$, but Δ_W is independent of which flow area is larger, as described later.

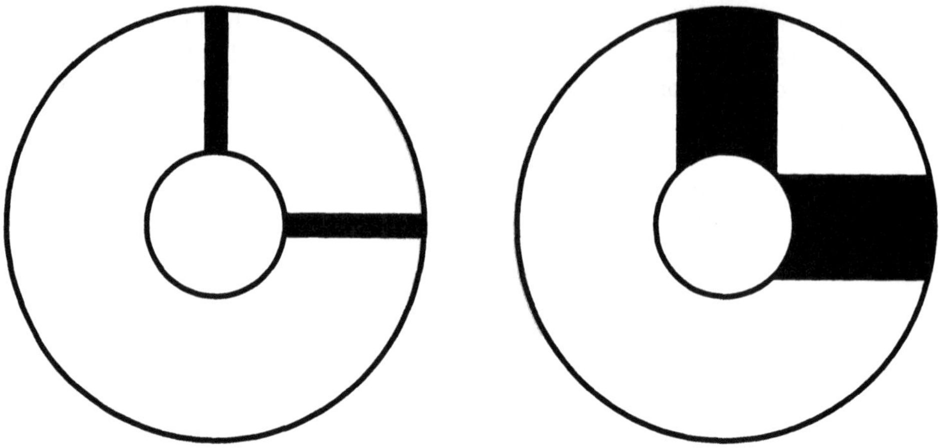

(a) (b)

Figure 6.43: Schematics of Two Example Regenerators with Extreme Configurations, Demonstrating the Effect of Varying w/R_o: (a) $w/R_o \to 0$ (b) $\phi(R_i/w) = \theta/2$.

nearly the same effectivenesses as otherwise-identical sector-seal regenerators for low w/R_o, but their effectivenesses decrease as w/R_o increases.

Along with w/R_o, the parameter R_i/R_o is also significant. In Figure 6.44 are represented two extreme cases. For case a, there should be little difference between the uniform-width-seal regenerator and an otherwise-identical one with sector-shape seals. This is because R_i/R_o is close to unity. For case b, the difference should be significant if R_i/R_o is sufficiently small. For such a regenerator, even if the seals are relatively thin, passages near the hub ($r \approx R_i$) spend a significantly smaller portion of the time exposed to the C_N-side flow, while passages near the rim ($r \approx R_o$) spend a significantly greater portion of the time exposed to the C_N-side flow. Again, the lower-effectiveness outer passages have more weight in determining the uniform-width-seal-regenerator effectiveness. Therefore, in general, uniform-width-seal regenerators have nearly the same effectivenesses as otherwise-identical sector-seal regenerators for high R_i/R_o, but their effectivenesses decrease with R_i/R_o.

Constraints on w/R_o and R_i/R_o. In performing ϵ_W calculations, constraints should be imposed on w/R_o and R_i/R_o. One obvious constraint is that R_i/R_o must be less than unity. Some other, perhaps less obvious constraints follow.

$$\frac{w/R_o}{R_i/R_o} \le 2 \tag{6.168}$$

$$\phi\left(\left[\frac{R_i}{R_o}\right]\left[\frac{R_o}{w}\right]\right) \le \frac{\theta}{2} \tag{6.169}$$

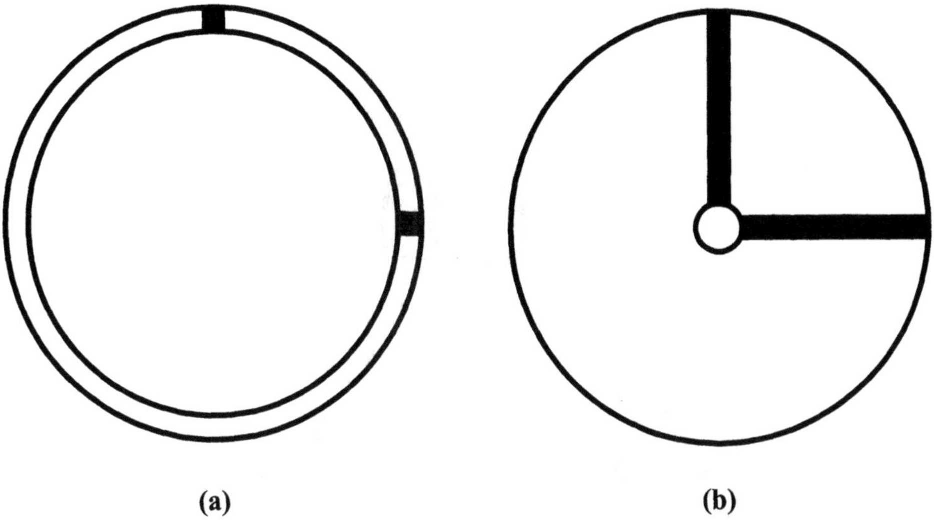

Figure 6.44: Schematics of Two Example Regenerators with Extreme R_i/R_o Values: (a) $R_i/R_o \to 1$ (b) $R_i/R_o \to 0$.

$$\phi\left(\left[\frac{R_i}{R_o}\right]\left[\frac{R_o}{w}\right]\right) \leq \pi - \frac{\theta}{2} \tag{6.170}$$

In practice, if any of these constraints were violated, a portion of the regenerator core would have zero effectiveness—some of the core would be useless as a heat exchanger. Example regenerators for which these constraints are violated are shown in Figure 6.45.

Δ_W vs. As'. Along with the two parameters w/R_o and R_i/R_o, which are discussed above, As' has a significant influence on the effectiveness of uniform-width-seal regenerators. Schematics of two regenerators with different As' values are shown in Figure 6.46. For case a, all passages spend roughly the same amount of time exposed to each flow if uniform-width seals are used. For case b, if the seals are closer together, passages near the hub spend a significantly smaller portion of the time exposed to the C_N-side flow, and passages near the rim spend a significantly larger portion of the time exposed to the C_N-side flow. In general, Δ_W increases in magnitude with the degree to which As' differs from unity.

Reductions in effectiveness should be independent of which flow-area is larger. Plots of Δ_W vs. w/R_o are shown in Figures 6.47–6.49 for $R_i/R_o = 0.1$, 0.3, 0.5, 0.7, and 0.9; and $As' = 0.1$, 0.5, and 1.0475. The parameter values, $NTU = 22$; $C_{RAT} = 1$; and $(hA)' = 1$, were chosen equal to those specified in the regenerator-design example provided by Wilson [73][15]. $C_{ROT} \to \infty$ was

[15] This was also the motive for choosing $As' = 1.0475$ instead of $As' = 1.0$. Wilson [73] found that $As' = 1.0475$ was nearly optimal for his particular regenerator-design example.

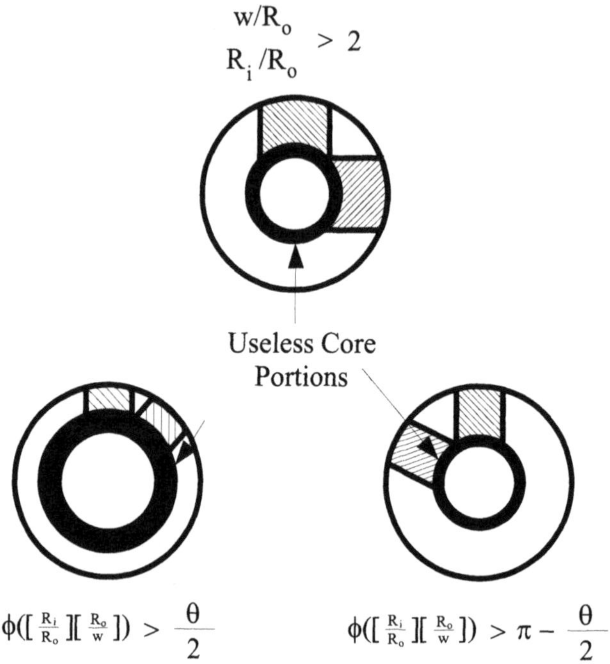

Figure 6.45: Schematics of Example Regenerators for which Constraints are Violated.

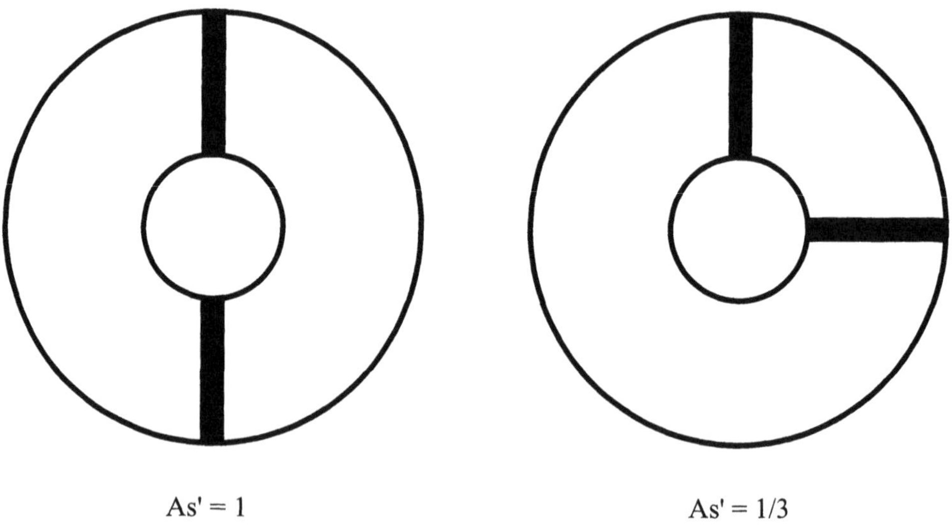

Figure 6.46: Schematics of Regenerators with Different Solid-Area Ratios.

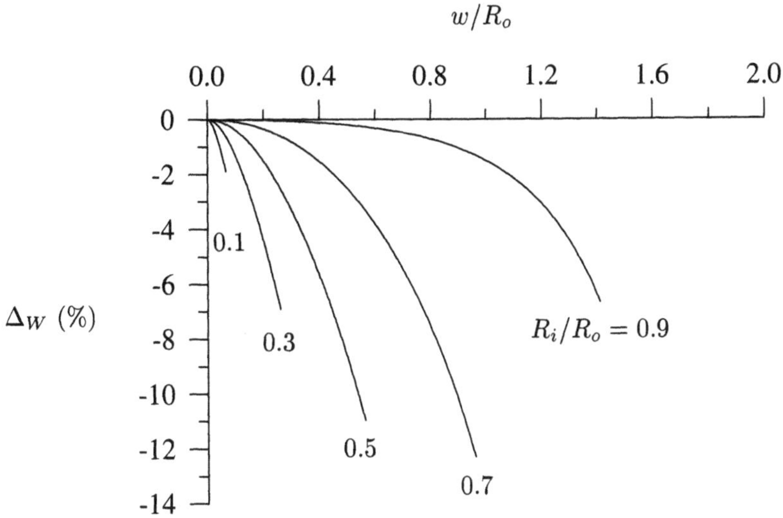

Figure 6.47: Plots of Δ_W vs. w/R_o for $NTU = 22$; $C_{RAT} = 1$; $(hA)' = 1$; $C_{ROT} \to \infty$; $As' = 0.1$; and $R_i/R_o = 0.1$, 0.3, 0.5, 0.7, and 0.9.

specified, as it was for the remainder of the given results (except those in the example calculation of ϵ_W for $C_{ROT} = 3$) to facilitate computation. $C_{ROT} \to \infty$ is a good approximation for determining effectiveness for sufficiently fast core rotation ($C_{ROT} \geq 7$). It is a fair approximation for somewhat slower rotation ($C_{ROT} \geq 3$). Identical results were obtained if identical seal locations were specified but the flows were switched (if $As' = 10$ instead of 0.1; if $As' = 2$ instead of 0.5; or if $As' = 0.95465$ instead of 1.0475)

The plots in Figures 6.47–6.49 show that for a given value of As', decreases in effectiveness incurred through the use of uniform-width seals increase in magnitude as w/R_o is increased or as R_i/R_o is decreased. Also, the more As' deviates from unity (the more unequal are the flow areas), the more effectiveness is reduced through the use of uniform-width seals.

Δ_W vs. NTU. Consider variation of the heat-transfer parameter NTU. The effects on reductions in effectiveness when uniform-width seals are used instead of sector seals are determined. Plots of ϵ'_W and ϵ vs. r/w, radial location, are shown in Figure 6.50 for $C_{RAT} = 1$; $(hA)' = 1$; $C_{ROT} \to \infty$; $As' = 0.95465$; $R_i/R_o = 0.5$; $w/R_o = 0.9$; and $NTU = 22$, 50, and 100. The plots of ϵ'_W start at a value higher than ϵ at radial locations near the hub ($r \approx R_i$) and drop below ϵ as the rim is approached ($r \approx R_o$). This feature is predicted above.

Another feature of the plots in Figure 6.50 is that the plot portions at low r/w values, associated with passage locations near the hub, flatten as NTU is increased. This is because the effectiveness of any heat exchanger is limited thermodynamically to unity. As NTU is increased, the effectivenesses of the

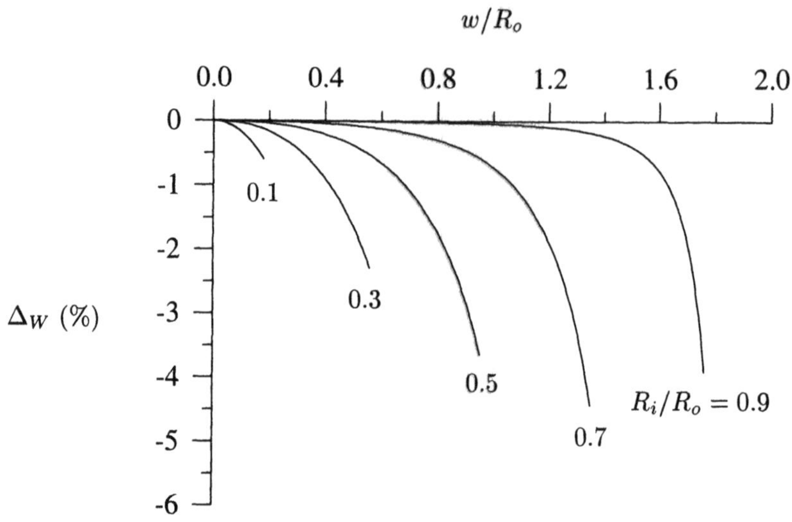

Figure 6.48: Plot of Δ_W vs. w/R_o for NTU = 22; C_{RAT} = 1; $(hA)'$ = 1; $C_{ROT} \to \infty$; $As' = 0.5$; and $R_i/R_o = 0.1$, 0.3, 0.5, 0.7, and 0.9.

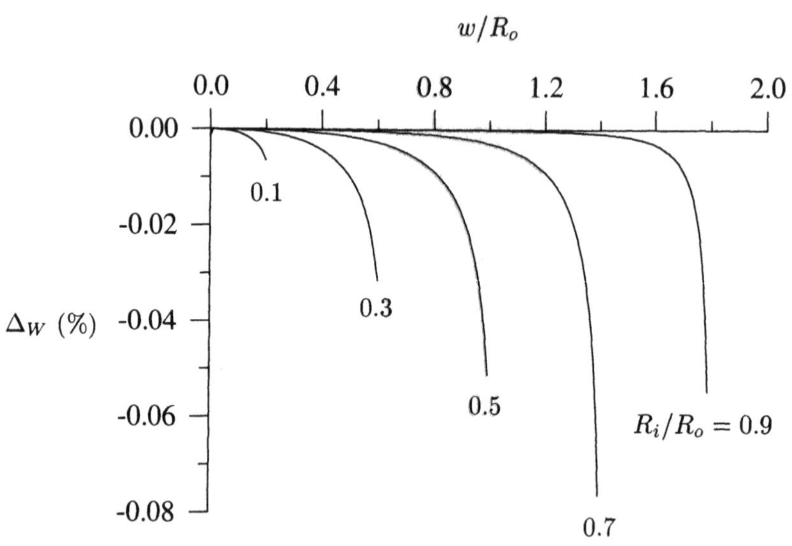

Figure 6.49: Plot of Δ_W vs. w/R_o for NTU = 22; C_{RAT} = 1; $(hA)'$ = 1; $C_{ROT} \to \infty$; $As' = 1.0475$; and $R_i/R_o = 0.1$, 0.3, 0.5, 0.7, and 0.9.

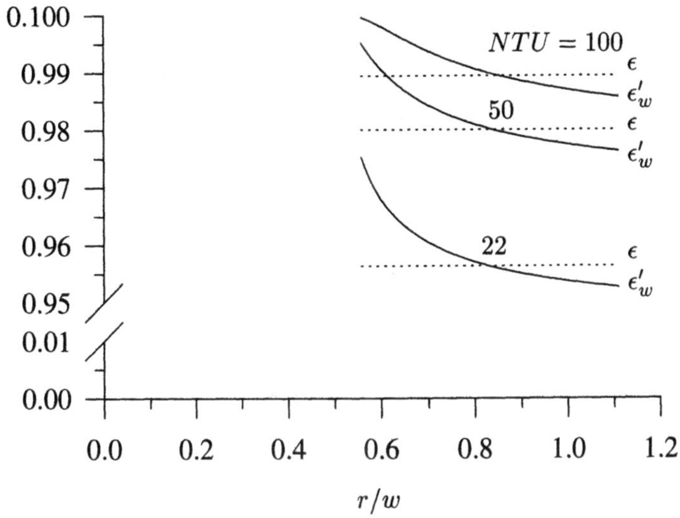

Figure 6.50: Plots of ϵ'_W and ϵ vs. r/w for $C_{RAT} = 1$; $(hA)' = 1$; $C_{ROT} \to \infty$; $As' = 0.95465$; $R_i/R_o = 0.5$; $w/R_o = 0.9$; and $NTU = 22$, 50, and 100.

passages near the hub increase more slowly as they approach unity than do the effectivenesses of passages at other locations. These effectivenesses of the passages near the hub do not offset the low effectivenesses of the passages near the rim. Thus, decreases in effectiveness incurred through use of uniform-width seals increase in magnitude with NTU. The plots of Δ_W vs. w/R_o for $C_{RAT} = 1$; $(hA)' = 1$; $As' = 1.0475$ or 0.95465; $R_i/R_o = 0.5$; and $NTU = 22$, 50, and 100, in Figure 6.51, prove this. For seal widths less than 30% of the rim radius, heat-transfer performance loss is negligible.

Δ_W vs. C_{RAT}. Now we consider variations of the heat-capacity-rate ratio, C_{RAT}. Figure 6.52 shows a plot of ϵ vs. C_{RAT} for $NTU = 22.0$; and $C_{ROT} \to \infty$. Clearly, $\partial\epsilon/\partial C_{RAT} < 0$. This conclusion is made by Lambertson [40] and is true in general. Also, the variation of ϵ_W with C_{RAT} increases with C_{RAT}: a change in C_{RAT} for high C_{RAT} produces a greater change in effectiveness than does a change in C_{RAT} for low C_{RAT}. This is also true in general.

In Figure 6.53 is shown a plot of $\Delta C_{RAT}/C_{RAT}$ vs. r/w for $NTU = 22$; $(hA)' = 1$; $C_{ROT} \to \infty$; $As' = 0.95465$; $R_i/R_o = 0.5$; and $w/R_o = 0.9$ where

$$\frac{\Delta C_{RAT}}{C_{RAT}} \equiv \frac{C_{RAT_W} - C_{RAT}}{C_{RAT}} . \tag{6.171}$$

The plot in Figure 6.53 is independent of C_{RAT}. The reason for this is clear from Equations 6.147 and 6.148. Therefore, the variations in C_{RAT} with radial location should be greatest for large C_{RAT}.

Effectiveness variations are greatest for C_{RAT} variations if C_{RAT} is large.

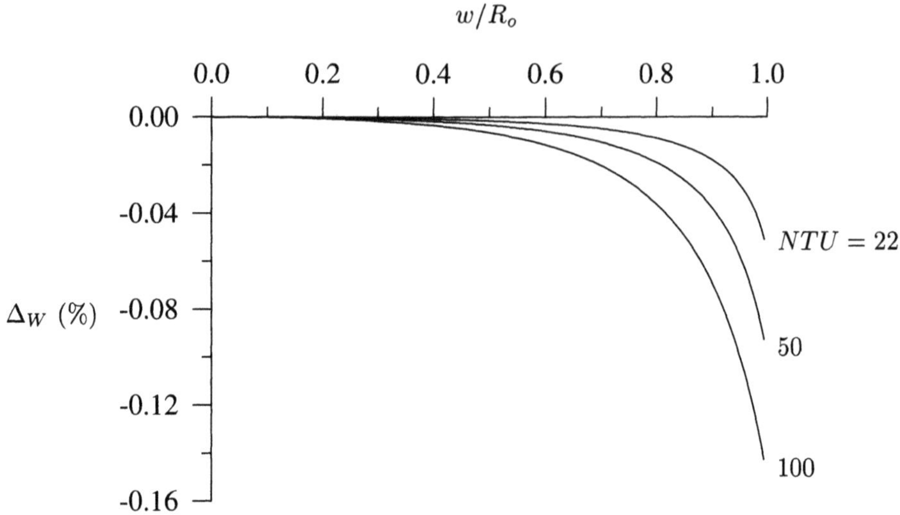

Figure 6.51: Plots of Δ_W vs. w/R_o for $C_{RAT} = 1$; $(hA)' = 1$; $As' = 1.0475$ or 0.95465; $R_i/R_o = 0.5$; and $NTU = 22$, 50, and 100.

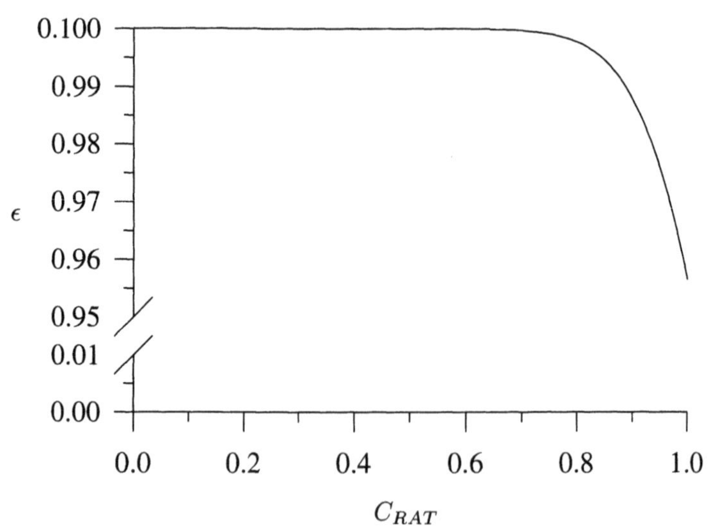

Figure 6.52: Plot of ϵ vs. C_{RAT} for $NTU = 22.0$; and $C_{ROT} \to \infty$

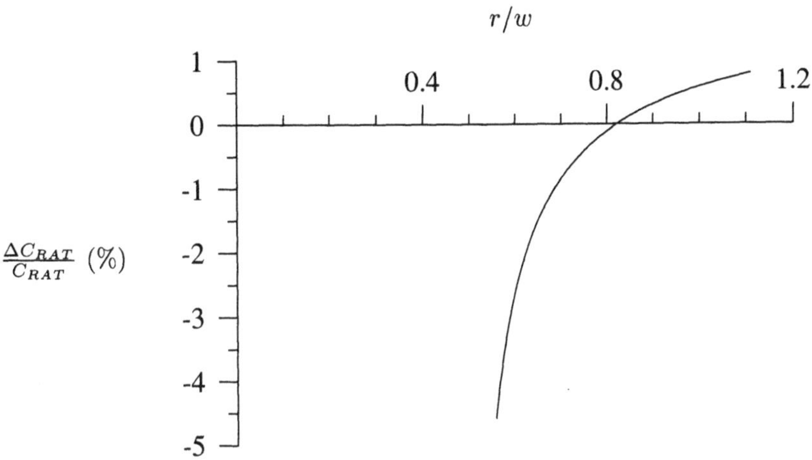

Figure 6.53: Plot of $\Delta C_{RAT}/C_{RAT}$ vs. r/w for $NTU = 22$; $(hA)' = 1$; $C_{ROT} \to \infty$; $As' = 0.95465$; $R_i/R_o = 0.5$; and $w/R_o = 0.9$.

Also, variations in C_{RAT} with radial location are greatest for large C_{RAT}. Therefore, effects of uniform-width seals are greatest for large C_{RAT}. The magnitude of the reductions in effectiveness incurred through the use of uniform-width seals increases with C_{RAT}. The plots of Δ_W vs. w/R_o in Figure 6.54 prove this.

Δ_W vs. $(hA)'$. Changes in Δ_W vs. w/R_o are minimal when $(hA)'$ is varied. The maximum amount of variation in ϵ that Lambertson [40] was able to achieve through varying $(hA)'$ was 7%. Therefore, slight variations of this parameter with radial location when uniform-width seals are used have little impact on the effectivenesses at those radial locations. Therefore, Δ_W is practically independent of $(hA)'$.

Δ_W vs. C_{ROT}. Determining the effects of C_{ROT} variation on Δ_W for large ranges of the other parameters would be an arduous task. A closed-form effectiveness expression exists only for $C_{ROT} \to \infty$: for finite C_{ROT}, effectiveness must be calculated numerically. For the Δ_W vs. w/R_o results given so far, w/R_o was incremented 200 times, while for each w/R_o value, r/w was incremented 100 times and ϵ'_w was calculated at each of these radial locations. Thus, for each Δ_W vs. w/R_o plot, an effectiveness computation was performed 20,000 times. These computations were completed in seconds on a VAXstation 3100. The program that the author uses to calculate effectiveness for finite C_{ROT} takes approximately one minute to run on a VAXstation 3100. Clearly, computation of Δ_W for large ranges of the other parameters is intractable.

Most optimal regenerative gas turbines have $C_{ROT} \geq 3$ (see Chapter 5).

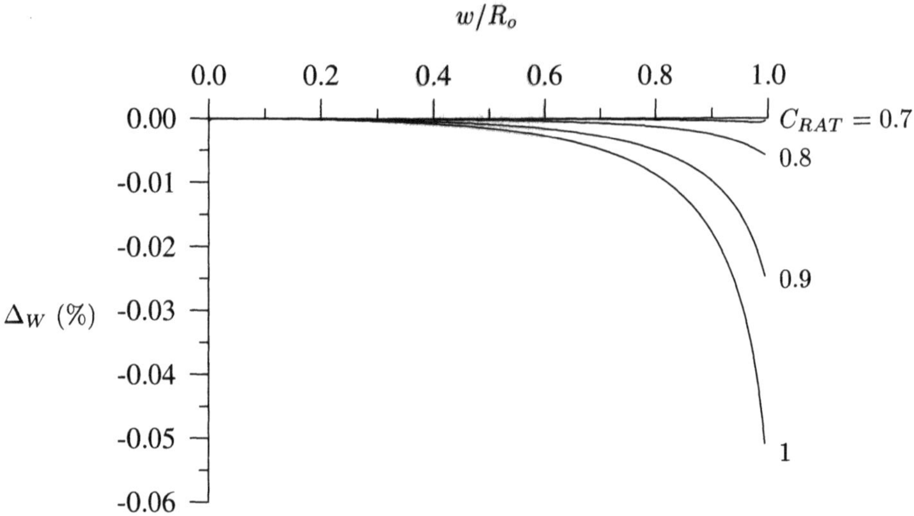

Figure 6.54: Plots of Δ_W vs. w/R_o for $NTU = 22$; $(hA)' = 1$; $C_{ROT} \to \infty$; $As' = 0.95465$; $R_i/R_o = 0.5$; and $C_{RAT} = 0.7$, 0.8, 0.9, and 1.

For such C_{ROT} values, effectiveness is close to that for $C_{ROT} \to \infty$. Therefore, results given in this section for $C_{ROT} \to \infty$ can be used for rough assessment of the reductions in effectiveness that would be incurred through the use of uniform-width seals instead of sector seals; however, the author recommends validation of a final design through calculation of ϵ_W[16]. An example of such a calculation is given next.

Example Calculation of ϵ_W for $C_{ROT} = 3$. Assume that a designer specifies the parameters $\epsilon = 0.95$; $C_{RAT} = 0.99$; $C_{ROT} = 3$; and $(hA)' = 1$ for a rotary-regenerator design. The designer then obtains $NTU = 20.975$ from a computer program incorporating Lambertson's [40] method, from charts provided by Kays and London [37] , or from some other source. Assume that the design process is continued, and $As' = 0.5$ and $R_i/R_o = 0.5$ are selected to achieve partial optimization according to some criteria. Assume that seal leakage would otherwise be great for this particular situation, so relatively thick seals are incorporated with $w/R_o = 0.6$. A schematic of the resulting regenerator is shown in Figure 6.55. Sector seals can be accomodated easily in a rotary-regenerator design process by simply increasing the core's face area (See Wilson [73]); however, for the actual regenerator, uniform-width seals would be used. It is of interest to determine the degree to which the assumption of sector seals would lead to reduction in the effectiveness of the final-design regenerator.

[16]Especially a design in which As' is significantly different from unity, and/or w/R_o is large, and/or R_i/R_o is small

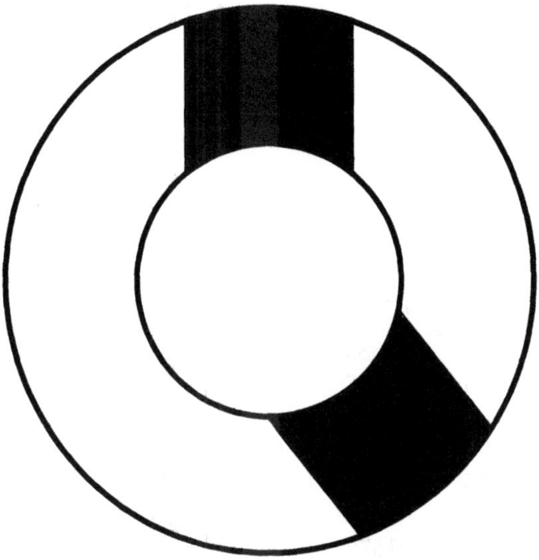

Figure 6.55: Schematic of a Regenerator with $As' = 0.5$; $R_i/R_o = 0.5$; and $w/R_o = 0.6$.

The reduction in effectiveness incurred through use of uniform-width seals can be estimated for this example from the plots given earlier. In this example, $(hA)' = 1$; $As' = 0.5$; $NTU = 20.975 \approx 22$; and $C_{RAT} = 0.99 \approx 1$. Figure 6.48, can be used if the small conservativism due to NTU actually being less than 22 (See Figure 6.51) and C_{RAT} actually being less than 1 (See Figure 6.53) is considered. The value for uniform-width-seal effect obtained from Figure 6.48 for $w/R_o = 0.6$ and $R_i/R_o = 0.9$ is $\Delta_W \approx -0.6\%$. The conservativism produces an actual value for this design example that is a bit less.

In Table 6.4 are shown intermediate results from a calculation using the a finite-difference integration scheme for this particular example. To obtain the ϵ_W value for this design example, the $\epsilon'_W \delta A_N / A_N$ values must be added to obtain $\epsilon_W = 0.9447$. This is associated with a uniform-width-seal effect of $\Delta_W = -0.56\%$. Therefore, for this design example, C_{ROT} is sufficiently large so that the results given in this section for $C_{ROT} \to \infty$ can be assumed to be accurate.

Conclusions Regarding Seal Shape. The following conclusions can be drawn from the analysis given in this section.

1. As', w/R_o, and R_i/R_o are principal variables on which reductions in rotary-regenerator effectiveness depend when uniform-width seals are used instead of sector seals for a particular regenerator design.

2. Reductions in effectiveness are most significant when the flow areas are highly unequal (As' is greatly different from unity). These reductions are

r/w	C'_{RAT_W}	C_{RAT_W}	$(hA)'_W$	NTU_W
0.87500	0.84689	0.84689	0.85544	22.609
0.95833	0.89467	0.89467	0.90371	22.036
1.04167	0.93062	0.93062	0.94002	21.623
1.12500	0.95875	0.95875	0.96844	21.311
1.20833	0.98140	0.98140	0.99132	21.066
1.29167	1.00006	0.99993	0.98994	20.870
1.37500	1.01572	0.98452	0.97468	21.031
1.45833	1.02905	0.97177	0.96205	21.167
1.54167	1.04055	0.96103	0.95142	21.282
1.62500	1.05056	0.95187	0.94235	21.381

C_{ROT_W}	$\delta A_N/A_N$	ϵ''_W	ϵ'_W	$\epsilon'_W \delta A_N/A_N$
4.0139	0.052318	0.99220	0.99220	0.051910
3.6393	0.063198	0.98368	0.98368	0.062167
3.3828	0.073903	0.97374	0.97374	0.071962
3.1955	0.084492	0.96361	0.96361	0.081417
3.0525	0.095002	0.95397	0.95397	0.090629
2.9398	0.10546	0.94518	0.94512	0.099672
2.8928	0.11587	0.95184	0.93711	0.10858
2.8528	0.12625	0.95693	0.92991	0.11740
2.8184	0.13661	0.96092	0.92348	0.12616
2.7883	0.14694	0.96410	0.91770	0.13485

Table 6.4: Intermediate Results for an Effectiveness Calculation for a Uniform-Width-Seal Regenerator with $NTU = 20.975$; $C_{RAT} = 0.99$; $(hA)' = 1$; $C_{ROT} = 3$; $As' = 0.5$; $R_i/R_o = 0.5$; and $w/R_o = 0.6$.

independent of which area is larger.

3. Reductions in effectiveness increase with seal thickness (w/R_o).

4. Reductions in effectiveness are greatest for regenerators with small core-hub diameters (small R_i/R_o).

5. Reductions increase with regenerator size, or NTU.

6. Reductions increase with heat-capacity-rate ratio (C_{RAT}).

7. Reductions are practically independent of convective-conductance ratio ($(hA)'$).

8. Rotation rate, for sufficiently fast rotation ($C_{ROT} \geq 3$), has a secondary effect on the reductions in effectiveness.

6.1.13 Discontinuous Rotation

This section describes the effects of discontinuous rotation on regenerator effectiveness. The principal conclusions are the following:

1. Discontinuous core rotation can reduce regenerator effectiveness.

2. Reductions in effectiveness due to discontinuous rotation can be completely avoided if the designer makes a good choice for the combination of seal location and discontinuous-rotation angle. Good choices are such that $180^o - \Delta$ is an even multiple of α_{ROT}/N_{ROT}, where: Figure 1.12 defines the seal-location angle, Δ; α_{ROT} is the discontinuous-rotation angle; and N_{ROT} is the number of times the core must rotate (rotating discontinuously) before is reaches its starting position. This number of rotations can be calculated by $N_{ROT} = LCM(360^o, \alpha_{ROT})/360^o$, where the function $LCM(360^o, \alpha_{ROT})$ is the Least Common Multiple of 360° and α_{ROT}.

The rest of this section gives analyses that lead to these conclusions. In the analyses we assume infinite core-rotation rate, zero axial conduction, and steady-state operation are.

Direct seal leakage in regenerators leads to decreased cycle thermal efficiency. A proposed solution to reduce direct seal leakage is discontinuous core rotation (see Wilson [75, 76] and Section 1.2.9). With discontinuous rotation, instead of moving at a constant angular velocity, the core moves quickly through a discontinuous-rotation angle, α_{ROT}, pauses, and then repeats this cycle. Clamping seals can be employed to eliminate direct seal leakage during pause periods. Thus, if the Rotation-period-to-Pause-period Ratio (RPR) is 10%, direct seal leakage could be reduced by approximately 91% (neglecting flow differences due to the faster moving core associated with discontinuous rotation).

This section addresses differences in effectiveness strictly due to discontinuous core rotation. First we show how effectiveness values calculated under the assumption of continuous rotation can be used for regenerators with discontinuous rotation. Also, we show why (for most cases) regenerator effectiveness for discontinuous rotation differs from that for continuous rotation. Then we quantify changes in effectiveness due to discontinuous rotation for various heat-transfer parameters. Finally we give conclusions and design guidelines.

Simplifying Assumptions. For analysis of discontinuous-rotation regenerators we make some simplifying assumptions. These assumptions are added to those made for analysis of continuous-rotation regenerators (see Section 6.1.3 for discussion of these assumptions). We assume that seals have negligible thickness. Direct seal leakage is high for thin seals. Therefore, in a practical setting seals must have a finite thickness. However, this thickness will not affect our basic conclusions.

Another assumption is an infinitely quick rotation, or $RPR = 0$. Practically, the rotation period would be limited by core stress-tolerance; however, the effects of a sufficiently small RPR would be negligible. Finally, differences in direct seal-leakage due to discontinuous rather than continuous rotation are not considered. Any increases in direct seal leakage during the rotation periods probably would not negate reductions during the pause periods.

Comparison of Effectiveness Calculations for Discontinuous-Rotation and Continuous-Rotation Regenerators. Consider the regenerators shown in Figures 1.2 and 6.56. The operation of the axial-flow rotary regenerator is discussed in Chapter 1. The regenerator in Figure 6.56 is a two-chamber, switching regenerator. For this kind, one side is heated while the other is cooled for a time period, τ_{switch}. At the end of this period, the flows are switched by the two-way valves, and the hot side is cooled and the cold side is heated. Then, the process is repeated.

It is specified that the cores and the associated flows in Figures 1.2 and 6.56 are identical from a heat-transfer standpoint (relevant parameters are equal). It is specified that the only difference between the two heat-exchanger configurations is that the core in Figure 1.2 rotates continuously with a rotation period, τ_{ROT}, and the core in Figure 6.56 is stationary and the flows are periodically switched. It also is specified that

$$\tau_{switch} = \frac{\tau_{ROT}}{2} \tag{6.172}$$

In the analysis given in this section, axial conduction is neglected. Therefore, effectiveness is not a function of As' for the analysis given in this section. This permits As' to be set arbitrarily while maintaining constant NTU, C_{RAT}, C_{ROT}, and $(hA)'$. For convenience, As' is set to $As' = 1$. Thus, in general the regenerator's seals in Figure 1.2 may divide the core in halves.

Regenerator effectiveness can be calculated from

$$\epsilon = \frac{T_{Xi} - \overline{T}_{Xo}}{C_{RAT}(T_{Xi} - T_{Ni})} \tag{6.173}$$

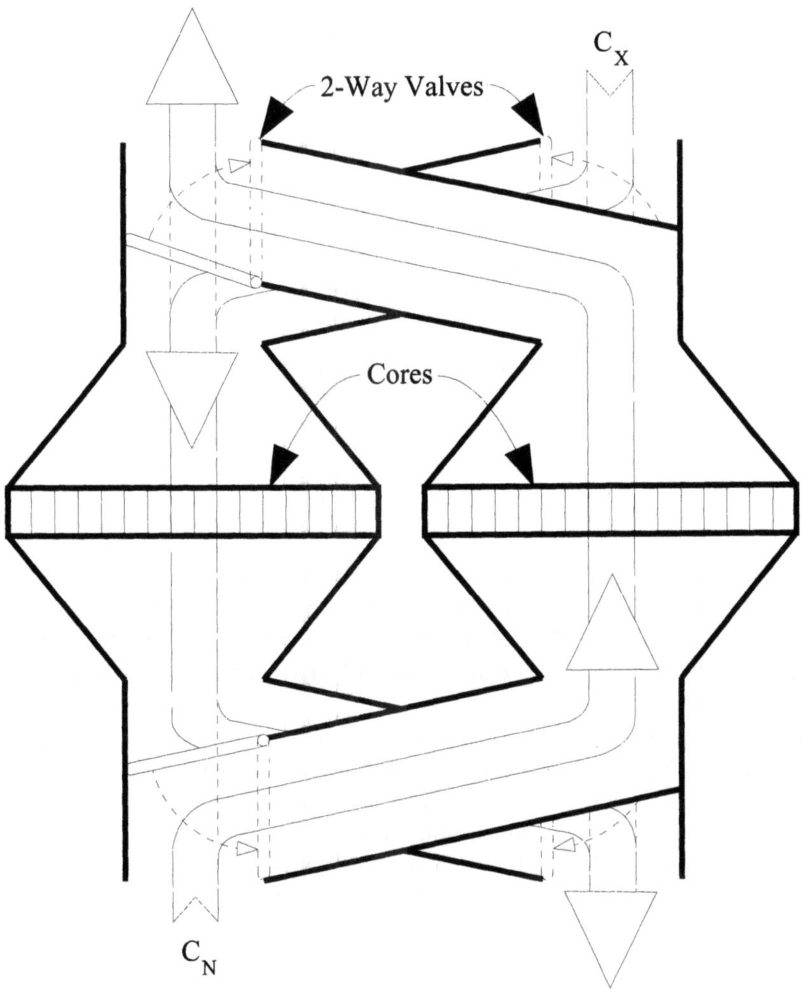

Figure 6.56: Schematic of a Two-Chamber Switching Regenerator.

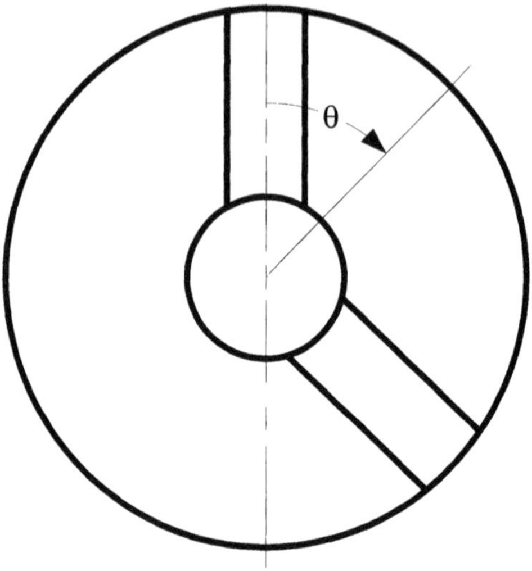

Figure 6.57: θ Definition.

or

$$\epsilon = \frac{\overline{T}_{No} - T_{Ni}}{T_{Xi} - T_{Ni}} \tag{6.174}$$

Fluid and core properties are assumed constant with temperature (see Section 6.1.3). Regenerator effectiveness is not a function of core-inlet temperatures (see Kays and London [37]). Therefore, in calculating a regenerator's effectiveness, core-inlet temperatures can be set to convenient values. If $T_{Ni} = 0$ and $T_{Xi} = 1$, then $\epsilon = \overline{T}_{No}$.

Consider the core passages of a continuous-rotation regenerator such as the one shown in Figure 1.2. Each one spends $\tau_{ROT}/2$ on each side. At any time, each passage on the C_N side will have spent a different amount of time on that side. Therefore, T_{No}, the core-outlet temperature of an individual passage on the C_N side, is different for each passage at each angle. The temperature, $\overline{T}_{No}$, is calculated by averaging the T_{No}'s over the total angle on the C_N side ($\pi \times As'/[1 + As']$).

$$\overline{T}_{No} = \int_0^1 T_{No}(\sigma) d\sigma \ , \tag{6.175}$$

where $\sigma = \theta/\pi$ and θ is defined in Figure 6.57.

Now consider the core passages of the regenerator shown in Figure 6.56. Each one alternately is exposed to each flow for $\tau_{switch} = \tau_{ROT}/2$. For this case, $\overline{T}_{No}$ is calculated by averaging the T_{No}'s over the period, τ_{switch}.

$$\overline{T}_{No} = \int_0^1 T_{No}(\gamma) d\gamma \tag{6.176}$$

where $\gamma = t_N/\tau_{switch}$ and t_N is the amount of time that the core portion has spent on the C_N side.

For a constant continuous-rotation rate,

$$\frac{\theta}{2\pi} = \frac{t_N}{\tau_{ROT}} \tag{6.177}$$

so that

$$\sigma = \frac{2t_N}{\tau_{ROT}} \tag{6.178}$$

and, from Equation 6.172

$$\sigma = \frac{t_N}{\tau_{switch}} \tag{6.179}$$

Therefore,

$$\sigma = \gamma \tag{6.180}$$

Therefore, switching regenerators and continuously-rotating regenerators with identical NTU, C_{RAT}, C_{ROT}, and $(hA)'$ values have identical effectivenesses, under the given assumptions.

Consider a continuous-rotation regenerator such as the one shown in Figure 1.2. If instead of rotating continuously, the core discontinuously rotates 180° every $\tau_{switch} = \tau_{ROT}/2$, the average outlet temperatures effected are identical to those of a switching regenerator (if the switching regenerator has identical dimensionless parameters to those of the continuous-rotation regenerator). Thus, as was just determined, the effectiveness is equal to that of the continuous-rotation regenerator.

Now consider what happens if the core discontinuously rotates an angle, α_{ROT} (degrees) every $\tau_{ROT}\alpha_{ROT}/360^o$ Figure[17] such that 360° is an even multiple of this angle (see Figure 1.12). To provide clarity, example regenerators are used throughout the rest of this section. For the example regenerator shown in Figure 6.58, $As' = 1$, $\alpha_{ROT} = 45^o$, and 360° is an even multiple of 45°. For this case, each 45° portion of the core spends an equal amount of time, $\tau_{ROT}/2$, on each side of the seals. Therefore, the effectiveness is equal to that of a continuous-rotation regenerator with identical NTU, C_{RAT}, C_{ROT}, and $(hA)'$ values.

Figure 1.12 shows the example regenerator core with an unequal area-split. The parameter, Δ is defined in the figure. Δ is related to As' by

$$As' = \frac{180^o - \Delta}{180^o + \Delta} \tag{6.181}$$

As discussed in above, we assume that effectiveness is not a function of As'. Therefore, provided the four dimensionless parameters of which effectiveness is a function are unchanged, the resulting regenerator again has an effectiveness equal to that of the associated continuous-rotation regenerator. For the example

[17] It is assumed throughout the rest of this section that the period of discontinuous rotation is $\tau_{disc} = \tau_{ROT}\alpha_{ROT}/360^o$.

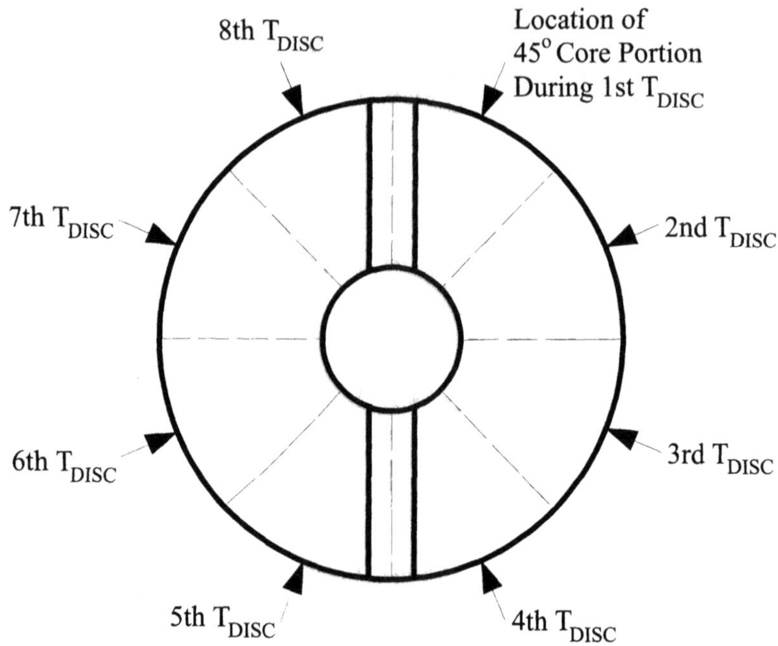

Figure 6.58: Example Regenerator Core with $\alpha_{ROT} = 45^o$ and $As' = 1$.

regenerator-core in Figure 1.12, each 45° core portion spends $3\tau_{ROT}/8$ seconds on the C_N side and $5\tau_{ROT}/8$ seconds on the C_X side. For the continuous-rotation case, the same is true of any core portion. In general, if $180^o - \Delta$ is an even multiple of α_{ROT}/N_{ROT} where $N_{ROT} \equiv LCM(360^o, \alpha_{ROT})/360^o$, and $LCM(x, y)$ is the least common multiple of x and y, then the discontinuous-rotation regenerator has an effectiveness equal to that of the continuous-rotation regenerator.

Now, consider the case where $180^o - \Delta$ is not an even multiple of α_{ROT}/N_{ROT}. In Figure 6.59 is shown the example regenerator's core with $\alpha_{ROT} = 45^o$ and $\Delta = 15^o$. Two portions of the example regenerator's core can be followed as the core rotates discontinuously. These portions are labeled 1 and 2 in Figure 6.59. Starting on the C_N side of the top seal, the portion locations during each discontinuous-rotation period are shown in Figure 6.59.

First, follow the 30° core portion 1. As the core rotates discontinuously, this portion alternately spends $4\tau_{disc}$ seconds on the C_N side and $4\tau_{disc}$ seconds on the C_X side. As for the 15° core portion 2, it alternately spends $3\tau_{disc}$ seconds on the C_N side and $5\tau_{disc}$ on the C_X side. Each 45° element of the core may be partitioned into similar 30° and 15° portions. Therefore, 2/3 of the example regenerator's core spends 4/8 of the time on the C_N side, and the remaining 1/3 of the example regenerator's core spends 3/8 of the time on the C_N side.

The example regenerator can be modelled as consisting of two separate sub-

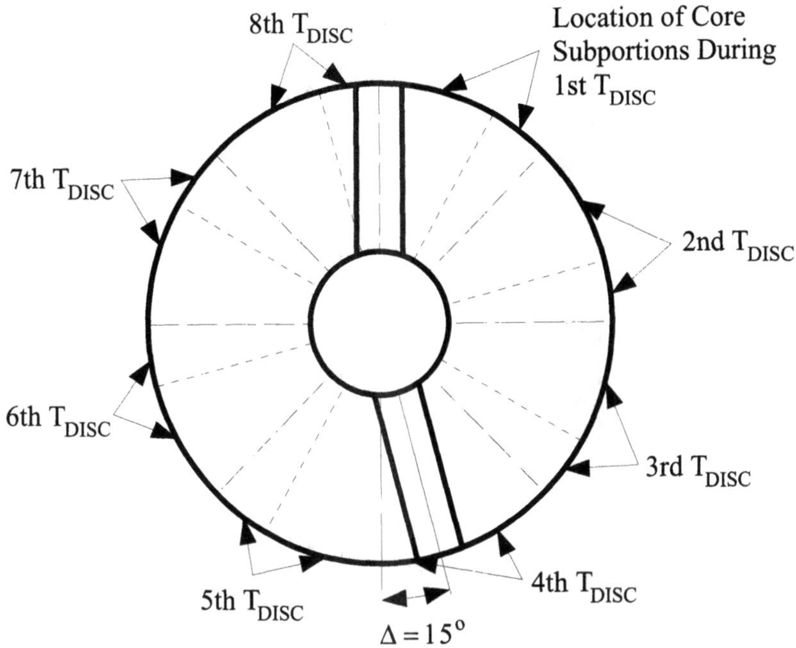

Figure 6.59: Example Regenerator Core with $\alpha_{ROT} = 45^o$ and $\Delta = 15^o$.

regenerators operating in parallel, as shown in Figure 6.60. Core 1 in the figure has a face area equal to 2/3 of the example regenerator core's face area, and it spends 4/8 of the time on the C_N side. Core 2 in the figure has a face area equal to 1/3 of the example regenerator core's face area, and it spends 3/8 of the time on the C_N side. Core 1 is associated with all of the 30° example regenerator's core portions that spend 4/8 of the time on the C_N side. Core 2 is associated with all of the 30° example regenerator's core portions that spend 3/8 of the time on the C_N side. The example regenerator can be modelled as having incoming fluid flows that are divided, 2/3 ducted to Core 1 and 1/3 ducted to Core 2. In this model, the outlet flows on the C_N-side are mixed, and the outlet flows on the C_X-side are mixed.

Now, it is shown how to calculate the model regenerator's effectiveness and, equivalently, that of the example regenerator. Note that Core 1 and Core 2 each have $\Delta_{1,2} = 45^o$. Therefore, the effectiveness of each of the two subregenerators is equal to that of a regenerator with an identical core that rotates continuously.

As described above, it may be specified that $T_{Xi} = 1$ and $T_{Ni} = 0$, so that the effectiveness of each subregenerator is just the fluid outlet-temperature on the C_N side of each subregenerator. The example regenerator's effectiveness is the average outlet temperature of the model regenerator after the two subregenerator's outlet-flows are mixed. Because 2/3 of the flow passes through Core 1 and 1/3 passes through Core 2, the example regenerator's effectiveness equals

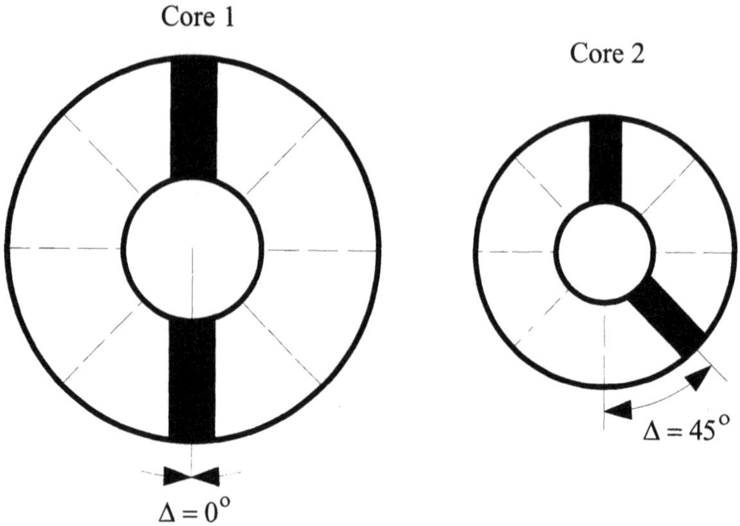

Figure 6.60: Example Regenerator Core Modelled as Two Separate Cores.

2/3 times the effectiveness of the subregenerator with Core 1 plus 1/3 times the effectiveness of the subregenerator with Core 2.

There is a caveat associated with this argument. Some combinations of Δ and α_{ROT} lead to cases where the C_{N_j}-side of one of the subregenerators is the C_X-side of the original regenerator and the C_{X_j}-side is the C_N-side of the original regenerator. For such cases, the model regenerator's effectiveness must be calculated differently. This issue is addressed shortly.

Two parameters are implicitly defined:

ζ_j (%) of the core spends χ_j (%) of the time on the C_N side

where j is equal to 1 or 2[18], depending upon which core portion is considered (core-portion 1 or core-portion 2 for the example core shown in Figure 6.59). Thus, for the example core, $\zeta_1 = 2/3$, $\chi_1 = 4/8$, $\zeta_2 = 1/3$, and $\chi_2 = 3/8$.

Calculation of Dimensionless Parameters from χ_j. Consider a core portion, j, of one of the two subregenerators. This core portion spends χ_j of the time on the C_N side. Thus,

$$\chi_j = \frac{As_{N_j}}{As_{N_j} + As_{X_j}} \tag{6.182}$$

or

$$\chi_j = \frac{As'_j}{1 + As'_j} \tag{6.183}$$

[18]If $180^o - \Delta$ is not an even multiple of α_{ROT}/N_{ROT}, then there are two and only two different amounts of time that any core portion can spend on the C_N side.

Solving for As'_j,

$$As'_j = \frac{\chi_j}{1-\chi_j} \tag{6.184}$$

As mentioned above, we assume that a regenerator's effectiveness is not a function of As', if the four dimensionless parameters of which effectiveness is a function are maintained. Discontinuous rotation, producing core portions with different As' values, now is shown to alter effectiveness.

The dimensionless parameter, C_{RAT}, can be expressed as

$$C_{RAT} = \left(\frac{c_{p_N}\rho_N U_N}{c_{p_X}\rho_X U_X}\right) As' \tag{6.185}$$

For each subregenerator,

$$C_{RAT_j} = \left(\frac{c_{p_N}\rho_N U_N}{c_{p_X}\rho_X U_X}\right) As'_j \tag{6.186}$$

because the values in parentheses do not change from the regenerator to each of the two subregenerators. Therefore,

$$C_{RAT_j} = C_{RAT}\frac{As'_j}{As'} \tag{6.187}$$

C_{RAT_j}, thus calculated, could have a value greater than one, depending upon As'_j and As'. By definition, this is impossible. Therefore, let us define

$$C'_{RAT_j} \equiv C_{RAT}\frac{As'_j}{As'} \tag{6.188}$$

and calculate C_{RAT_j} as follows:

$$C_{RAT_j} = \begin{cases} C'_{RAT_j} & \text{if } C'_{RAT_j} \leq 1 \\ \frac{1}{C'_{RAT_j}} & \text{if } C'_{RAT_j} > 1 \end{cases} \tag{6.189}$$

Physically, if $C'_{RAT_j} > 1$, then the associated subregenerator has a C_{N_j}-side that is the C_X-side and a C_{X_j}-side that is the C_N-side: the original regenerator's sides of minimum and maximum heat capacity are reversed for subregenerator j.

Now, consider the dimensionless parameter, $(hA)'$. This parameter can be expressed as

$$(hA)' = \frac{h_N A_N}{h_X A_X} \tag{6.190}$$

or

$$(hA)' = \frac{h_N}{h_X}As' \tag{6.191}$$

where h is a convective-heat-transfer coefficient. For each of the subregenerators,

$$(hA)'_j = \begin{cases} \frac{h_N}{h_X}As'_j & \text{if } C'_{RAT_j} \leq 1 \\ \frac{h_X}{h_N As'_j} & \text{if } C'_{RAT_j} > 1 \end{cases} \tag{6.192}$$

because the heat-transfer coefficents, h_N and h_X, for the subregenerators are identical to those of the original regenerator. Thus,

$$(hA)'_j = \begin{cases} (hA)'\frac{As'_j}{As'} & \text{if } C'_{RAT_j} \leq 1 \\ \frac{As'}{(hA)'As'_j} & \text{if } C'_{RAT_j} > 1 \end{cases} \tag{6.193}$$

Now consider NTU. This parameter can be expressed as

$$NTU = \left[\frac{(hA)_N}{C_N}\right]\left[\frac{1}{1+(hA)'}\right] \tag{6.194}$$

Similarly, NTU_j can be expressed as

$$NTU_j = \begin{cases} \left[\frac{h_N(\zeta_j A_N)}{c_{pN}(\zeta_j \dot{m}_N)}\right]\left[\frac{1}{1+(hA)'_j}\right] & \text{if } C'_{RAT_j} \leq 1 \\ \left[\frac{h_X(\zeta_j A_X)}{c_{pX}(\zeta_j \dot{m}_X)}\right]\left[\frac{1}{1+(hA)'_j}\right] & \text{if } C'_{RAT_j} > 1 \end{cases} \tag{6.195}$$

because both A_N and $\dot{m}_N$ or A_X and $\dot{m}_X$ are reduced by the factor ζ_j for each of the subregenerators. The ζ_j's cancel, so

$$NTU_j = \begin{cases} NTU\left[\frac{1+(hA)'}{1+(hA)'_j}\right] & \text{if } C'_{RAT_j} \leq 1 \\ NTU\left[\frac{C_{RAT}}{(hA)'}\right]\left[\frac{1+(hA)'}{1+(hA)'_j}\right] & \text{if } C'_{RAT_j} > 1 \end{cases} \tag{6.196}$$

The fourth dimensionless parameter, C_{ROT}, can be expressed as

$$C_{ROT} = \frac{\dot{m}_R c_R}{\dot{m}_N c_{pN}} \tag{6.197}$$

For the subregenerators,

$$C_{ROT_j} = \begin{cases} \frac{(\zeta_j \dot{m}_R)c_R}{(\zeta_j \dot{m}_N)c_{pN}} & \text{if } C'_{RAT_j} \leq 1 \\ \frac{(\zeta_j \dot{m}_R)c_R}{(\zeta_j \dot{m}_X)c_{pX}} & \text{if } C'_{RAT_j} > 1 \end{cases} \tag{6.198}$$

because both $\dot{m}_R$ and $\dot{m}_N$ or $\dot{m}_X$ are reduced by the factor ζ_j. The ζ_j's cancel, so

$$C_{ROT_j} = \begin{cases} C_{ROT} & \text{if } C'_{RAT_j} \leq 1 \\ C_{ROT}C_{RAT} & \text{if } C'_{RAT_j} > 1 \end{cases} \tag{6.199}$$

Calculation of Discontinuous-Rotation-Regenerator Effectiveness. Using the dimensionless parameters developed in the last section, the effectiveness for each of the two subregenerators, ϵ_1 and ϵ_2 can be calculated. If $C'_{RAT_j} > 1$ for either of the subregenerators, then $T_{Xi_j} = T_{Ni}$ and $T_{Ni_j} = T_{Xi}$. Therefore, if $T_{Xi} = 1$, $T_{Ni} = 0$, and $C'_{RAT_j} > 1$, then $T_{Xi_j} = 0$ and $T_{Ni_j} = 1$. The C_N-side outlet flow from this subregenerator has an average temperature,

$\epsilon_j C_{RAT_j}$. The regenerator's effectiveness is then the weighted average of the C_N-side outlet-flow average temperatures.

$$\epsilon = \sum_{j=1}^{2} \zeta_j \times \begin{cases} \epsilon_j & \text{if } C'_{RAT_j} \le 1 \\ \epsilon_j C_{RAT_j} & \text{if } C'_{RAT_j} > 1 \end{cases} \tag{6.200}$$

The effectiveness for the example regenerator shown in Figure 6.59 is calculated now. Set $C_{RAT} = (hA)' = 1$ and $NTU = 20$. Using Equations 6.184, 6.188, 6.189, 6.193, 6.196, and 6.199 and $\chi_1 = 4/8$ and $\chi_2 = 3/8$, the following dimensionless parameters for the two subregenerators are obtained.

$$\begin{aligned} C'_{RAT_1} &= 1.182\ ; \\ C_{RAT_1} &= 0.8462\ ; \\ (hA)'_1 &= 0.8462\ ; \\ NTU_1 &= 21.67\ ;\ \text{and} \\ C_{ROT} &= \infty\ . \end{aligned}$$

and

$$\begin{aligned} C'_{RAT_2} &= 0.7091\ ; \\ C_{RAT_2} &= 0.7091\ ; \\ (hA)'_2 &= 0.7091\ ; \\ NTU_2 &= 23.40\ ;\ \text{and} \\ C_{ROT} &= \infty\ . \end{aligned}$$

Using Equation 6.1, we obtain the following effectiveness values for each of the two subregenerators:

$$\epsilon_1 = 0.9943 \tag{6.201}$$

and

$$\epsilon_2 = 0.9997\ . \tag{6.202}$$

Using Equation 6.200 and $\zeta_1 = 2/3$ and $\zeta_2 = 1/3$, the effectiveness of the discontinuous-rotation regenerator is obtained.

$$\epsilon_{disc} = 0.8941 \tag{6.203}$$

The effectiveness of the continuous-rotation regenerator can be calculated from Equation 6.1.

$$\epsilon_{cont} = 0.9524 \tag{6.204}$$

For this example, the effectiveness of the continuous-rotation regenerator with $As' = (180^o - 15^o)/(180^o + 15^o) = 0.8462$ decreases by 6.121% when it is rotated discontinuously through the discontinuous-rotation angle, $\alpha_{ROT} = 45^o$.

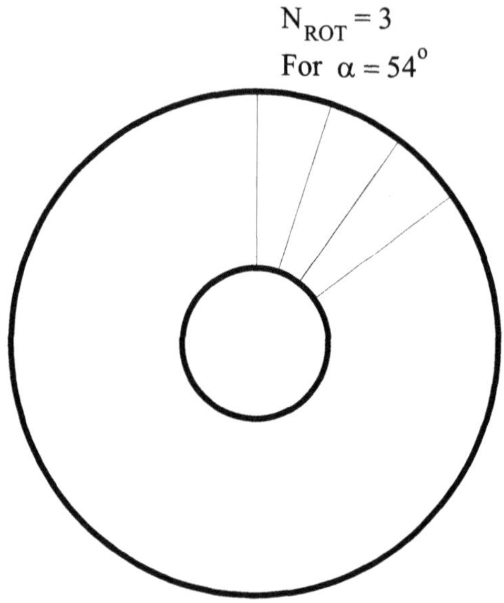

Figure 6.61: Division of α_{ROT} Core-Portion into N_{ROT} Subportions for a Discontinuous-Rotation Regenerator with $\alpha_{ROT} = 54^o$ and $\Delta = 6^o$.

Method for Determining ζ_j and χ_j. From the analysis above, it is clear that the ζ_j and the χ_j are the parameters that determine effectiveness changes due to discontinuous rotation. A method for determining these parameters is described now.

The first step is to divide into N_{ROT} subportions the core portion of angular extent α_{ROT} immediately to the C_N side of the C_N-side entry seal. This step is shown in Figure 6.61 for a discontinuous-rotation regenerator with $\alpha_{ROT} = 54^o$ and $\Delta = 6^o$. The subportion immediately to the C_N side of the C_N-side entry seal is then advanced N times, until the C_X-side entry seal is straddled, as shown in Figure 6.62.

Values for the ζ_j and χ_j can then be determined. The ζ_j values are determined by examining the subportion when it straddles the C_X-side entry-seal. The ζ_1 value is the fraction of the subportion's area on the C_N side of the seal. The ζ_2 value is the fraction of the subportion's area on the C_X side of the seal, or $\zeta_2 = 1 - \zeta_1$. The χ_j values are calculated as follows.

$$\chi_1 = \frac{\alpha_{ROT}(N \times N_{ROT})}{360^o N_{ROT}} \qquad (6.205)$$

$$\chi_2 = \frac{\alpha_{ROT}(N \times N_{ROT} - 1)}{360^o N_{ROT}} \qquad (6.206)$$

$$(6.207)$$

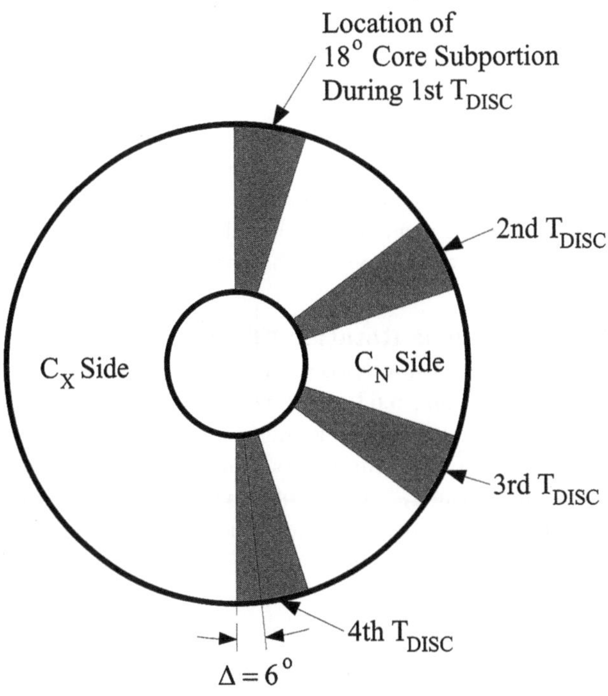

Figure 6.62: Propagation of Core Subportion for a Discontinuous-Rotation Regenerator with $\alpha_{ROT} = 54^o$ and $\Delta = 6^o$.

Changes in Effectiveness Due to Discontinuous Rotation. Discontinuous-rotation regenerator effectiveness was calculated above. It was shown that discontinuous rotation can produce effectivenesses that are different from those produced under continuous rotation. It was shown that differences in effectiveness are due to four parameters, χ_1, χ_2, ζ_1, and ζ_2. These parameters are functions of two other parameters, Δ and α_{ROT}.

Now it is shown how the effectiveness of a discontinuous-rotation regenerator varies with Δ and α_{ROT} for various system parameter values. Throughout the given analysis, $C_{ROT} = \infty$ is used for simplicity. This is the only C_{ROT} value for which an analytical effectiveness solution exists (see Section 6.1.5). In practice, regenerator cores that rotate sufficiently fast approach this limit. The purpose of this analysis is to give qualitative trends and obtain design guidelines. Therefore, the added difficulty associated with obtaining effectiveness values numerically for slow core-rotation rates is not warranted.

A regenerator with a discontinuous-rotation angle, $\alpha_{ROT} = 120^o$, will be considered. A discontinuous-rotation effect will be defined that quantifies changes in effectiveness due to discontinuous rotation. Variations of this discontinuous-rotation effect with NTU, C_{RAT}, and $(hA)'$ will be considered. Also, discontinuous-rotation-effect variations with α_{ROT} will be considered.

Variation of Discontinuous-Rotation Effect with NTU for $\alpha_{ROT} = 120^o$. The core of a regenerator with a discontinuous-rotation angle, $\alpha_{ROT} = 120^o$, is shown schematically in Figure 6.63 along with continuous-rotation subregenerators that mimic its behavior. As Δ is increased from $\Delta = -60^o$ to $\Delta = 60^o$, the size of Core 1 decreases and the size of Core 2 increases. The relative sizes of these cores determine the ζ_j. A plot of ζ_j vs. $-60^o \leq \Delta \leq 60^o$ is shown in Figure 6.64[19]. Similar weightings occur for other α_{ROT}. For all α_{ROT}, ζ_1 starts at 1 and decreases linearly to 0, and ζ_2 starts at 0 and increases linearly to 1, maintaining the sum $\zeta_1 + \zeta_2 = 1$.

$NTU = 0.1$. Consider a regenerator with $\alpha_{ROT} = 120^o$, $C_{RAT} = (hA)' = 1$, and $NTU = 0.1$. In Figure 6.65 is shown a plot of the C_N-side outlet temperatures vs. Δ for each of the subregenerators associated with this regenerator. Notice that the outlet temperature of Core 1 starts at the outlet temperature of a continuous-rotation regenerator with $C_{RAT} = (hA)' = 1$ and $NTU = 0.1$, where this outlet temperature has a weighting of 1. From this point, this outlet temperature decreases with its weighting. Effectiveness increases with decreasing C_{RAT_j} (see Section 6.1.5). The As' value decreases as Δ increases, so C_{RAT_1} decreases ($C'_{RAT_j} > 1$) to produce a higher effectiveness. The decreasing C_{RAT_1} weighting ($T_{Xo_j} = \epsilon_1 C_{RAT_1}$ for $T_{Xi} = 1$ and $T_{Ni} = 0$) produces a decreasing C_N-side outlet temperature for Core 1.

[19]Discontinuous-rotation regenerator cores with $|\Delta| > 180^o - \alpha_{ROT}$ are not considered in this analysis. Such regenerators have increasingly poorer effectiveness as $|\Delta|$ increases, because a portion of the core spends either no time on the C_X side ($\Delta < \alpha_{ROT} - 180^o$) or no time on the C_N side ($\Delta > 180^o - \alpha_{ROT}$).

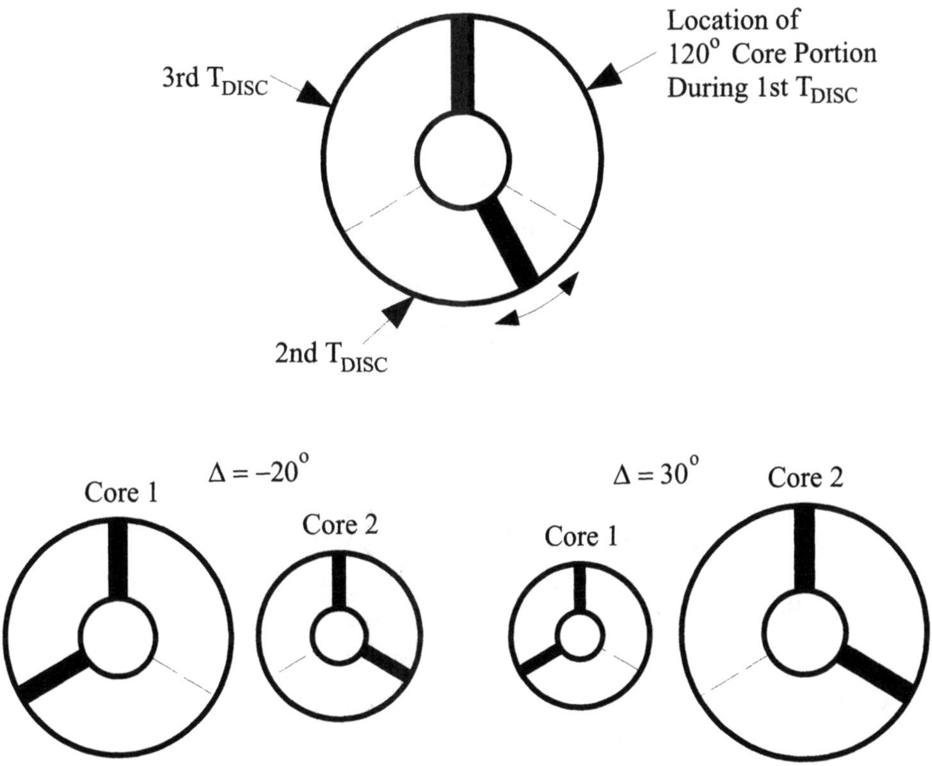

Figure 6.63: Schematic of a Regenerator Core with a Discontinuous-Rotation Angle, $\alpha_{ROT} = 120^o$ and Subregenerator Cores for $\Delta = -20^o$ and $\Delta = 30^o$.

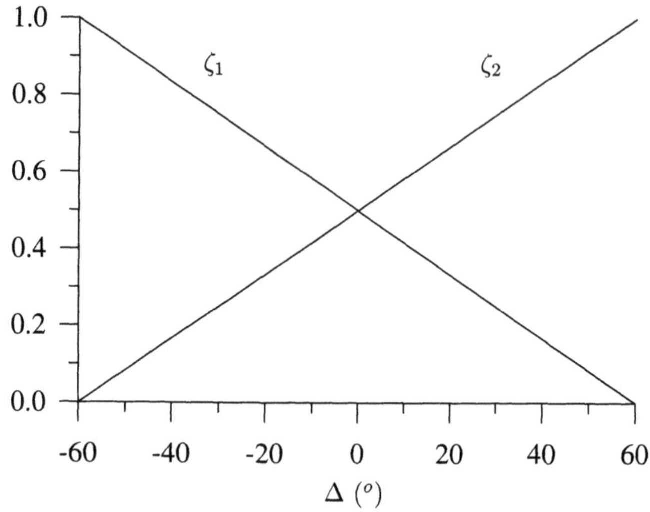

Figure 6.64: ζ_j vs. $-60^o \leq \Delta \leq 60^o$ for $\alpha_{ROT} = 120^o$.

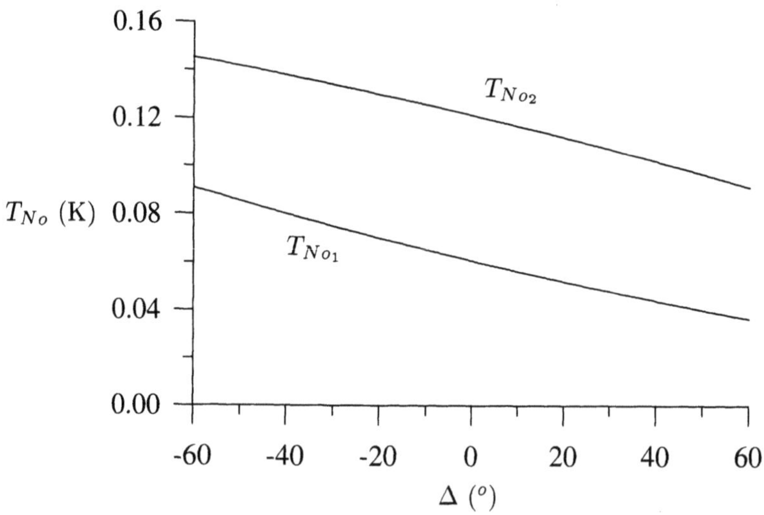

Figure 6.65: C_N-Side Outlet Temperatures vs. Δ for Each Subregenerator of a Discontinuous-Rotation Regenerator with $\alpha_{ROT} = 120^o$, $C_{RAT} = (hA)' = 1$, and $NTU = 0.1$.

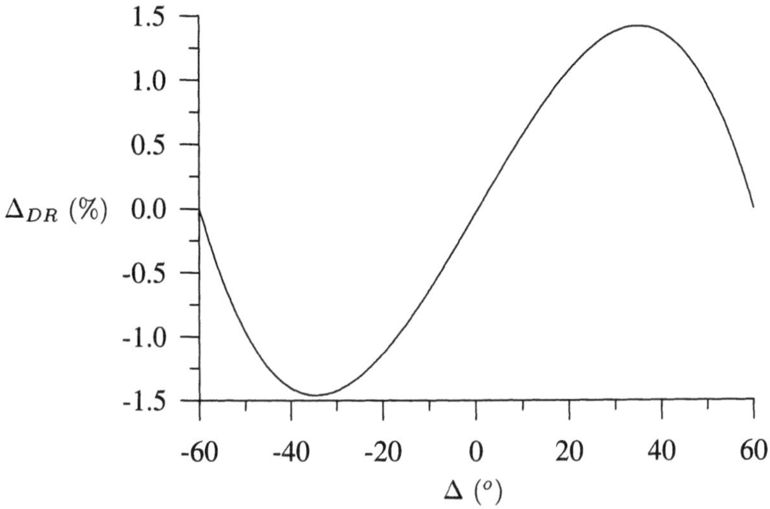

Figure 6.66: Δ_{DR} vs. Δ for a Discontinuous-Rotation Regenerator with $\alpha_{ROT} = 120^o$, $C_{RAT} = (hA)' = 1$, and $NTU = 0.1$.

Now consider the Core 2 C_N-side outlet temperature. This outlet temperature starts at a high value for $\Delta = -60^o$, where it has zero weighting. From this point, it drops to the outlet temperature of a continuous-rotation regenerator with $C_{RAT} = (hA)' = 1$ and $NTU = 0.1$, where it has a weighting of 1. The reason for this drop is that C_{RAT_2} increases to 1. This decreases this subregenerator's effectiveness.

A discontinuous-rotation effect is defined as follows.

$$\Delta_{DR} \equiv \frac{\epsilon_{disc} - \epsilon_{cont}}{\epsilon_{cont}} \tag{6.208}$$

Such notation is similar to that used by Bahnke and Howard [2]. In Figure 6.66 is shown a plot of Δ_{DR} vs. Δ. The large weightings on the low C_N-side outlet temperatures of Core 1 produce the downward-curving sinus for $\Delta < 0$, while the large weightings on the high C_N-side outlet temperatures of Core 2 produce the upward-curving sinus for $\Delta > 0$.

$NTU = 10.0$. A regenerator with $\alpha_{ROT} = 120^o$; $C_{RAT} = (hA)' = 1$; and $NTU = 10.0$ is now considered. A plot of C_N-side outlet temperatures for each of its two subregenerators is shown in Figure 6.67. Notice that the outlet temperature graph for Core 1 is of a similar form to that for the regenerator with $NTU = 0.1$, which was considered previously. The outlet temperature graph for Core 2 is quite different. This is because a regenerator's effectiveness has an upper bound of 1. Effectiveness increases with NTU (see Equation 6.1), so for high NTU values, the C_N-side outlet temperatures of Core 2 are limited.

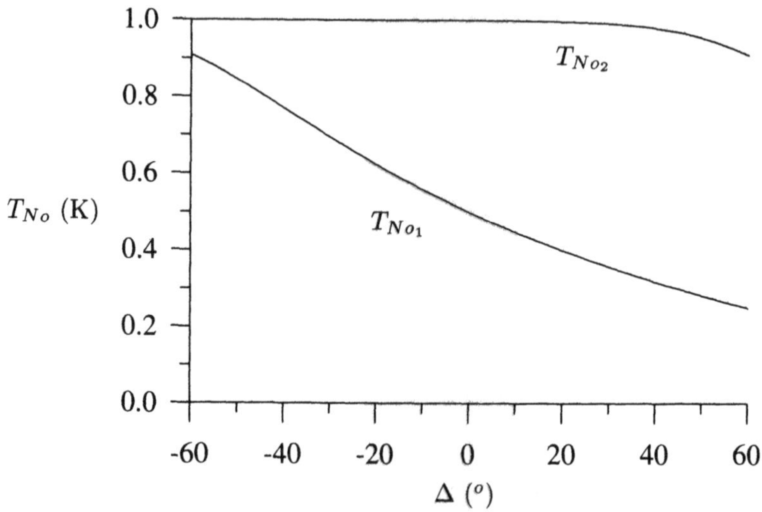

Figure 6.67: C_N-Side Outlet Temperatures for Each Subregenerator of a Discontinuous-Rotation Regenerator with $\alpha_{ROT} = 120^o$; $C_{RAT} = (hA)' = 1$; and $NTU = 10$.

A plot of Δ_{DR} vs. Δ is shown in Figure 6.68. The C_N-side outlet temperature of Core 2 is limited, so the downward-curving sinus dips deeper. Also, the upward-curving sinus is greatly reduced in magnitude.

Generalizations can be made now. Increasing NTU produces discontinuous-rotation regenerators with greater effectiveness reductions from effectivenesses of identical continuous-rotation regenerators. Also, these decreases occur over an increasingly larger Δ-range. Finally, Δ ranges of effectiveness increase diminish with increasing NTU.

Variation of Discontinuous-Rotation Effect with C_{RAT} for $\alpha_{ROT} = 120^o$. The variation of Δ_{DR} with C_{RAT} must be considered. In Figure 6.69 are shown plots of C_N-side outlet temperatures vs. Δ for discontinuous-rotation regenerators with $\alpha_{ROT} = 120^o$; $(hA)' = 1$; $NTU = 20$; and $C_{RAT} = 1$ and $C_{RAT} = 0.7$. For the regenerator with $C_{RAT} = 0.7$, $C'_{RAT_1} < 1$ for $\Delta < 180(1 - 2C_{RAT})/(1 + 2C_{RAT})$. Therefore, the C_N-side outlet temperature of Core 1 is not weighted by C_{RAT_1} in this Δ-range. For $\Delta > 180(1 - 2C_{RAT})/(1 + 2C_{RAT})$, $C'_{RAT_1} > 1$, and the remainder of the graph is similar to that for $C_{RAT} = 1$.

In Figure 6.70 are shown plots of Δ_{DR} vs. Δ for various C_{RAT} values. The graphs decrease in magnitude and shift to the right for decreasing C_{RAT}.

For small NTU and $C_{RAT} \neq 1$, Equation 6.1 reduces to

$$\epsilon \approx \frac{NTU}{1 + C_{RAT} NTU} \tag{6.209}$$

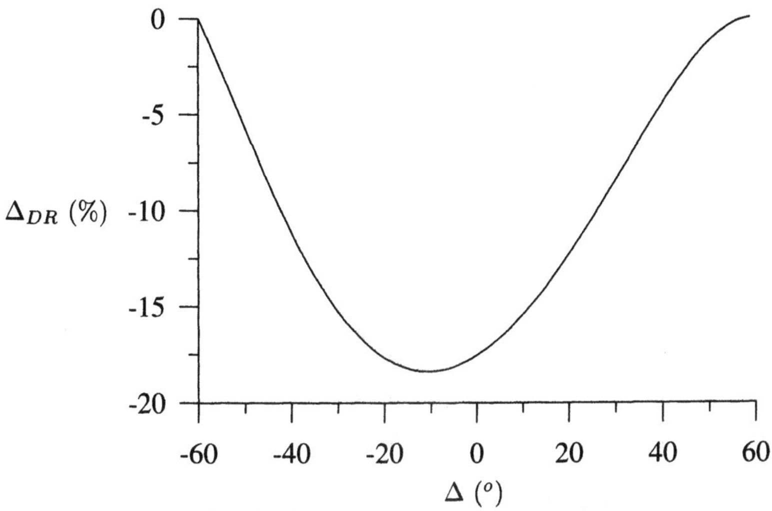

Figure 6.68: Δ_{DR} vs. Δ for a Discontinuous-Rotation Regenerator with $\alpha_{ROT} = 120^o$; $C_{RAT} = (hA)' = 1$; and $NTU = 10$.

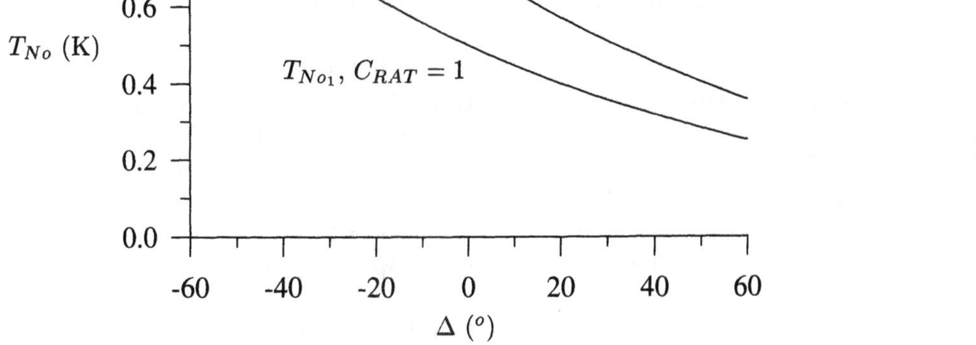

Figure 6.69: C_N-Side Outlet Temperatures vs. Δ for Discontinuous-Rotation Regenerators with $\alpha_{ROT} = 120^o$; $(hA)' = 1$; $NTU = 20$; and $C_{RAT} = 1$ and $C_{RAT} = 0.7$.

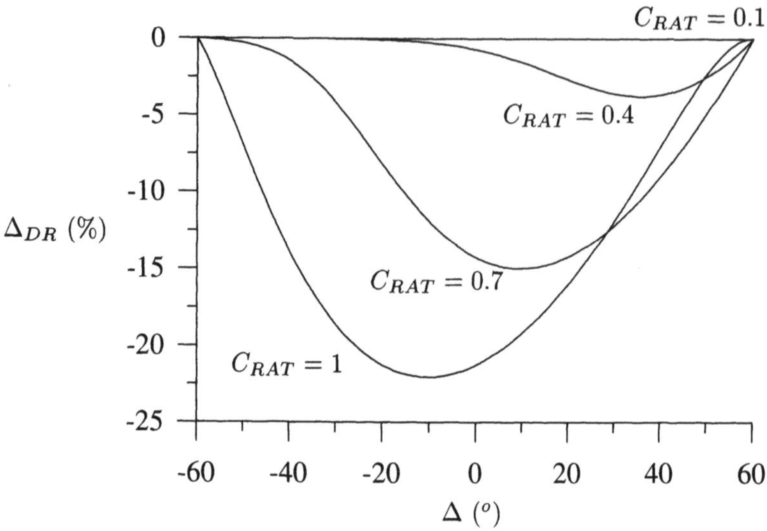

Figure 6.70: Δ_{DR} vs. $-60 < \Delta < 60$ for Discontinuous-Rotation Regenerators with $\alpha_{ROT} = 120^o$; $(hA)' = 1$; $NTU = 20$; and Various C_{RAT} Values.

For small NTU, Equations 6.209 and 6.1 reduce to

$$\epsilon \approx NTU \tag{6.210}$$

Thus, for small NTU, effectiveness is not a function of C_{RAT}. As NTU and effectiveness decrease, C_{RAT} effects become negligible. Plots of Δ_{DR} vs. Δ for discontinuous-regenerators with $\alpha_{ROT} = 120^o$, $(hA)' = 1$, $NTU = 0.1$, and $C_{RAT} = 1$ and $C_{RAT} \to 0$ are shown in Figure 6.71. Clearly the effects of C_{RAT} on Δ_{DR} (and therefore regenerator effectiveness) become negligible for small NTU. Also, the effects of varying C_{RAT} on discontinuous-rotation effect become negligible for small NTU.

Variation of Discontinuous-Rotation Effect with $(hA)'$ for $\alpha_{ROT} = 120^o$. The effects on Δ_{DR} of varying $(hA)'$ must be considered. In Figures 6.72, 6.73, and 6.74 are shown plots of Δ_{DR} vs. Δ for discontinuous-rotation regenerators with $\alpha_{ROT} = 120^o$; $C_{RAT} = 1$; $(hA)' = 0$, 1, and ∞; and $NTU = 10$, 5, and 1, respectively. For relatively large NTU ($NTU > 5$ is most common in gas-turbine applications), effects of varying $(hA)'$ are clearly negligible.

Variation of Discontinuous-Rotation Effect with α_{ROT}. The trends just described were drawn from consideration of discontinuous-rotation regenerators with discontinuous-rotation angles, $\alpha_{ROT} = 120^o$. Now, these trends are extended to discontinuous-rotation regenerators with arbitrary α_{ROT}.

Plots of Δ_{DR} vs. Δ for discontinuous-rotation regenerators with $C_{RAT} = (hA)' = 1$; $NTU = 10$ and 25; and $\alpha = 45^o$, 30°, and 20° are shown in

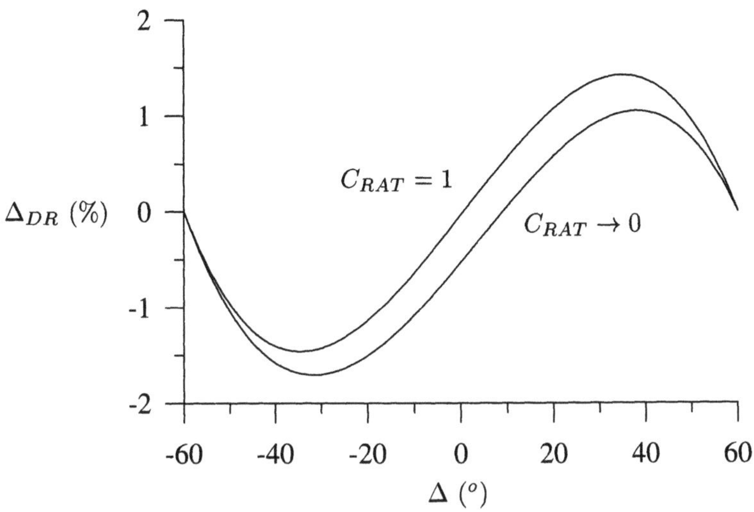

Figure 6.71: Δ_{DR} vs. Δ for Discontinuous-Rotation Regenerators with $\alpha_{ROT} = 120^o$; $(hA)' = 1$; $NTU = 0.1$; and $C_{RAT} = 1$ and $C_{RAT} \to 0$.

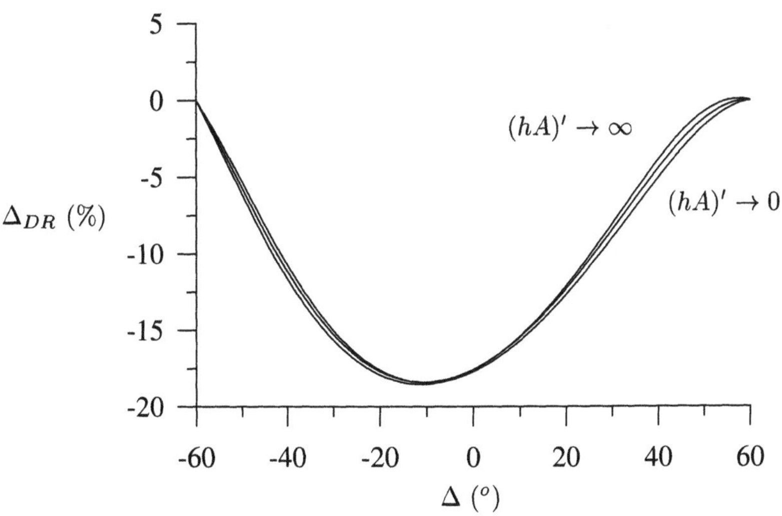

Figure 6.72: Δ_{DR} vs. Δ for Discontinuous-Rotation Regenerators with $\alpha_{ROT} = 120^o$; $C_{RAT} = 1$; $(hA)' = 0$, 1, and ∞; and $NTU = 10$.

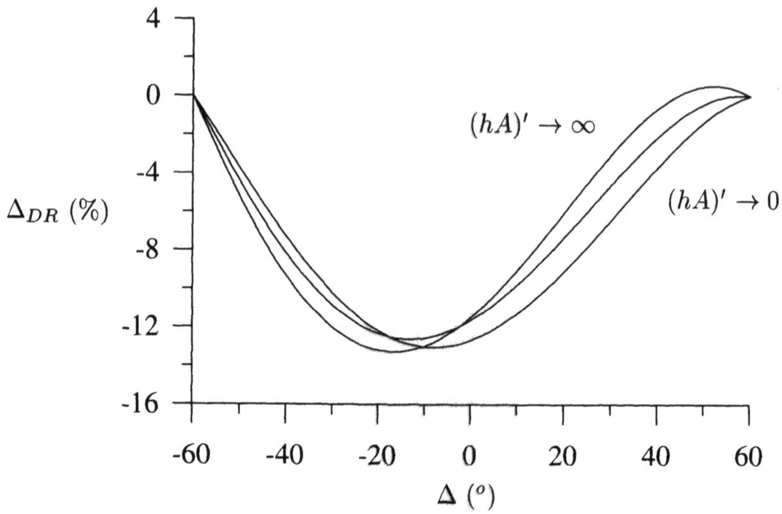

Figure 6.73: Δ_{DR} vs. Δ for Discontinuous-Rotation Regenerators with $\alpha_{ROT} = 120^o$; $C_{RAT} = 1$; $(hA)' = 0$, 1, and ∞; and $NTU = 5$.

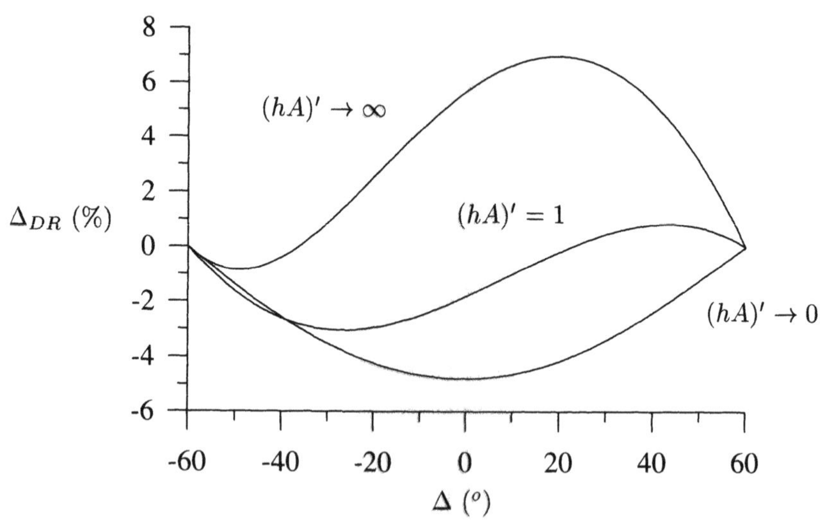

Figure 6.74: Δ_{DR} vs. Δ for a Discontinuous-Rotation Regenerator with $\alpha_{ROT} = 120^o$; $C_{RAT} = 1$; $(hA)' = 0$, 1, and ∞; and $NTU = 1$.

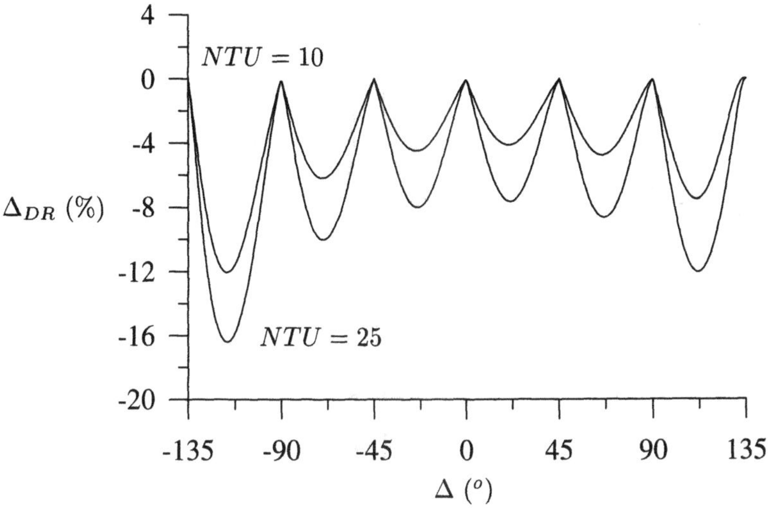

Figure 6.75: Δ_{DR} vs. Δ for Discontinuous-Rotation Regenerators with $C_{RAT} = (hA)' = 1$; $NTU = 10$ and 25; and $\alpha_{ROT} = 45^o$.

Figures 6.75, 6.76, and 6.77, respectively. The sinuses in the graphs will be referred to as humps. These plots show the following trends.

1. Regions of increasing effectiveness vanish for large NTU values and the associated large effectiveness values.
2. The sizes of humps increase with NTU.
3. The sizes of the humps decrease with the number of humps.

Trend 1 indicates that for large NTU, the designer needs to be concerned with detrimental effects of discontinuous-rotation. Trend 2 indicates that the effects of discontinuous-rotation become increasingly more pronounced with increasing NTU. Finally, Trend 3 indicates that the designer should, if possible, increase the number of humps to decrease the detrimental effects of discontinuous rotation.

For large NTU, the number of humps, N_h, in a Δ_{DR} vs. $\alpha_{ROT} - 180^o < \Delta < 180^o - \alpha_{ROT}$ plot is

$$N_h = N_{ROT}\left(\frac{360^o}{\alpha_{ROT}} - 2\right) \qquad (6.211)$$

To maximize the number of humps, the designer should choose α_{ROT} such that N_{ROT} is large. Some angles which achieve this are $\alpha_{ROT} = 19^o$, 23°, 26°, 29°, and 31°.

The detrimental effects of discontinuous-rotation are nonlinear in α_{ROT}. For $\alpha_{ROT} = 18^o$, $N_h = 18$, while for $\alpha_{ROT} = 20^o$, $N_h = 16$. Both of these

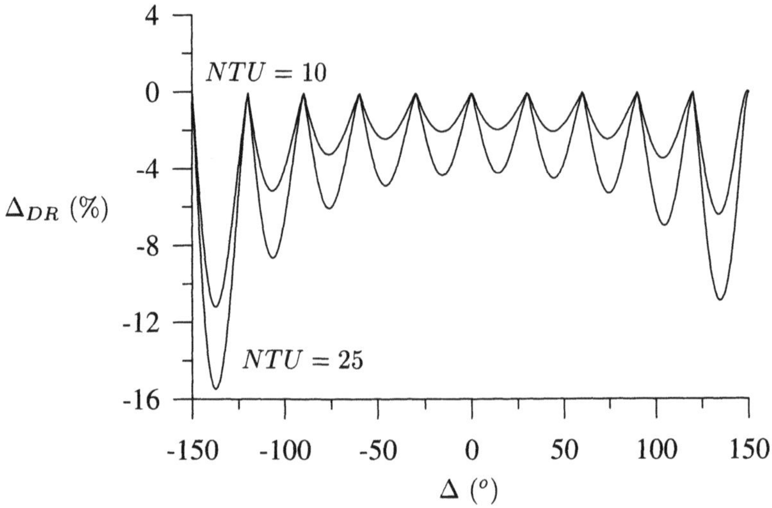

Figure 6.76: Δ_{DR} vs. Δ for Discontinuous-Rotation Regenerators with $C_{RAT} = (hA)' = 1$; $NTU = 10$ and 25; and $\alpha_{ROT} = 30^o$.

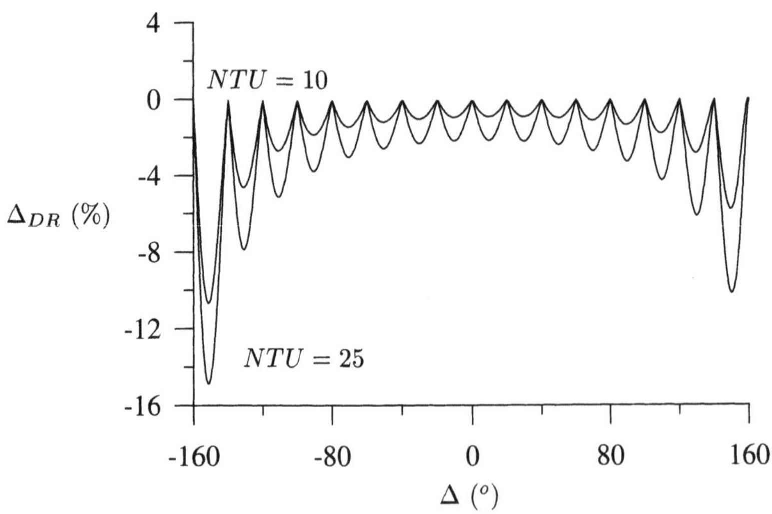

Figure 6.77: Δ_{DR} vs. Δ for Discontinuous-Rotation Regenerators with $C_{RAT} = (hA)' = 1$; $NTU = 10$ and 25; and $\alpha_{ROT} = 20^o$.

discontinuous-rotation angles could cause significant effectiveness reductions for certain seal locations. For $\alpha_{ROT} = 19^o$, $N_h = 322$. For this angle, which is between the other two, negligible effectiveness decreases would be incurred through discontinuous rotation.

Conclusions. From the analysis given in this section, the following conclusions can be drawn.

1. For small NTU values, effectiveness can increase through discontinuous rotation.

2. For large NTU values, appropriate for gas-turbine applications, effectiveness can only decrease through discontinuous-rotation.

3. Increases in NTU increase the detrimental effects of discontinuous-rotation on regenerator effectiveness.

4. Decreases in C_{RAT} reduce the detrimental effects.

5. Variation of $(hA)'$ has little effect for large NTU values.

6. Some discontinuous-rotation angles produce negligible changes in effectiveness, but other angles produce extreme changes.

Design Guidelines. From these conclusions, some design guidelines can be proposed. For low-effectiveness ($\epsilon < 0.5$) regenerators, the designer should try to exploit possible increases in effectiveness through discontinuous rotation. For high-effectiveness regenerators, the designer should avoid the detrimental effects incurred through discontinuous rotation by appropriate choice of the rotation angle.

To avoid the detrimental effects incurred through discontinuous rotation, the designer should choose discontinous-rotation angles for which changes in effectiveness due to discontinuous-rotation are negligible for all seal locations. Some such angles are listed in Table 6.5. If the designer must choose an angle for which an effectiveness decrease could occur through discontinuous rotation, this decrease in effectiveness still will be avoided if the discontinuous-rotation angle, α_{ROT}, is chosen such that $180^o - \Delta$ is an even multiple of α_{ROT}/N_{ROT}.

6.2 Summary

This section has focused on the heat-transfer performance of regenerators. We have given results of analyses and experimentation with regenerator cores.

First, the physical phenomena in regenerators were discussed in Section 6.1.2. Then, simplifications for the analysis were made in Section 6.1.3 and governing equations were derived in Section 6.1.4. In Section 6.1.5, the effectiveness of regenerators with very fast core-rotation rates was calculated. Finite core-rotation rates were considered in Section 6.1.6. In Section 6.1.7, axial conduction during

α_{ROT}
7°
11°
13°
14°
17°
19°
21°
22°
23°
26°
29°
31°
34°
37°
39°
41°
43°
47°
49°

Table 6.5: Discontinuous-Rotation Angles for which Changes in Effectiveness Due to Discontinuous-Rotation are Negligible for All Seal Locations.

flow exposures was considered. Heat diffusion when passages are under seals was included in Section 6.1.8. In Section 6.1.10, the effects of low core porosity on parallel-plate-regenerator effectiveness was addressed. In Section 6.1.11, the responses of regenerators to cold start-ups were considered. The effects on effectiveness of two causes of flow non-uniformity, face-seal shape and discontinuous rotation, were considered in Sections 6.1.12 and 6.1.13.

The next two sections address two additional facets of regenerator performance, leakage and pressure drops.

6.3 Seal Leakage

Harper [26] gives a comprehensive analysis of coupled direct and carry-over seal leakages. The principal result of his analysis is the following implicit expression for the total seal leakage (direct and carry-over combined), which is plotted in Figure 6.78:

$$\frac{1}{1+\frac{\dot{m}_{XLC}}{\dot{m}_{XL\Sigma}}} - \frac{1}{1+\left(\frac{\dot{m}_{XLC}}{\dot{m}_{XL\Sigma}}\right)\left(\frac{1}{r_R}\right)} - \ln\left[\frac{1-\left(\frac{\dot{m}_{XLC}}{\dot{m}_{XL\Sigma}}\right)\left(\frac{1}{r_R}\right)}{1-\frac{\dot{m}_{XLC}}{\dot{m}_{Xl\Sigma}}}\right] = \left(\frac{BS}{A}\right)^2 , \qquad (6.212)$$

where

$\dot{m}_{XL\Sigma}$	=	Total seal leakage (kg/s), direct and carry-over combined;
$\dot{m}_{XLC}$	=	Carry-over leakage (kg/s);
r_R	=	Pressure ratio across the seals (–); and
BS/A	=	Dimensionless core-rotation speed (–).

Equation 6.212 applies for each of the two seals separately:

1. Positive-rotation seal (the seal under which the core rotates from the compressed-air side to the exhaust side); and

2. Negative-rotation seal (the seal under which the core rotates from the exhaust side to the compressed-air side).

Harper [26] found that the direct and carry-over leakages are coupled in one direction: direct leakage depends on carry-over leakage, but carry-over leakage occurs independent of direct leakage. Carry-over leakage can be calculated directly:

$$\dot{m}_{XLC} = C_{ROT}\left[\frac{p}{1-p}\right]\left[\frac{1}{(\rho c)_R}\right] C_N \rho_c \, , \qquad (6.213)$$

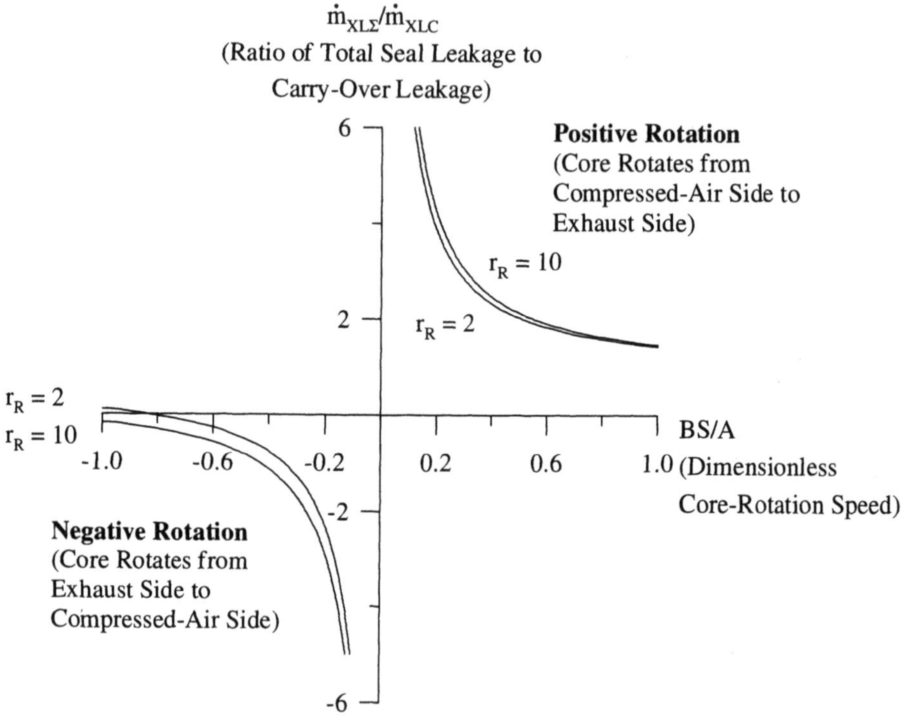

Figure 6.78: Seal Leakage Calculated Using the Method of Harper [20].

where

$\dot{m}_{XLC}$	=	Carry-over leakage (kg/s);
C_{ROT}	=	Dimensionless core rotation rate (–);
p	=	Core porosity (–), the ratio of voids volume to total volume;
$(\rho c)_R$	=	Specific heat capacity per volume of the core material (J/m^3-K);
C_N	=	Heat-capacity rate of the compressed air (W/K); and
ρ_c	=	Density of the gas that is carried over (kg/m^3).

After calculating carry-over leakage, we use Harper's [26] method to calculate the total leakage (see Section 5.1 for details on calculating total seal leakge using Harper's method). Figure 6.78 shows how total seal leakage varies with the dimensionless core rotation speed and the pressure ratio across the seals. First consider positive rotation. For slow core rotation most of the leakage is direct seal leakage, so the ratio of the total leakage to the carry-over leakage is high. For faster core rotation, carry-over leakage is a larger percentage of the total. For any given core rotation speed, total leakage is greatest for large pressure ratios across the seals.

Now consider negative rotation. For negative rotation, carry-over leakage is always negative: exhaust is always carried over from the exhaust side to the compressed-air side. For sufficiently slow core rotation rates, the total leakage is positive, so the ratio of total leakage to carry-over leakage is usually negative. The exception is when the core rotates very fast and carry-over leakage becomes greater than direct leakage. For this situation the net leakage is from the exhaust side to the compressed-air side (negative total leakage).

This section has considered seal leakage in regenerators. The next section considers pressure drops across the core and across the inlet and outlet headers.

6.4 Pressure Drops

The last two sections have considered heat transfer and seal leakage in regenerators. This section considers a third facet of regenerator performance: the pressure drops across the core and across the inlet and outlet headers.

In a regenerator, a serious effectiveness reduction can be incurred through a non-uniform flow distribution in the core (see Wilson [72]). The flow distribution can be made uniform by properly shaping the inlet and outlet headers; however, these headers will have finite pressure drops, which will reduce cycle performance. London, et al. [44] present analyses of flows in two-dimensional oblique-flow headers that effect uniform flow distributions in the core. Their results can be used to calculate the pressure drop across each side (the exhaust and compressed-air sides) of a regenerator. Their results for *two-dimensional* headers can be applied to calculate pressure drops in *three-dimensional* headers for design work (see Chapter 5), because the functional dependencies of the relevant parameters are similar for three-dimensional oblique-flow headers, and

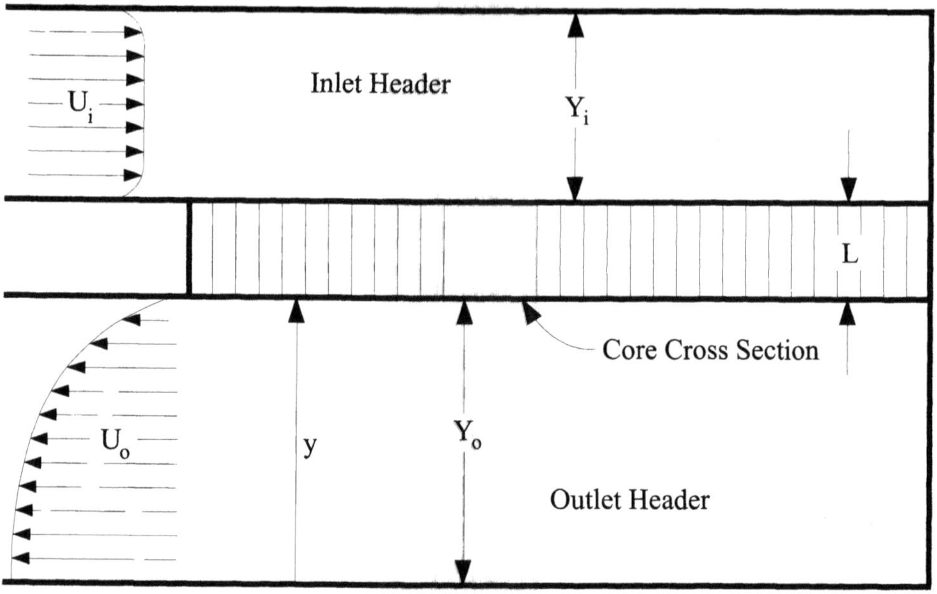

Figure 6.79: Schematic of a Regenerator-Core Cross Section with Inlet and Outlet Headers.

the magnitudes of the pressure drops of three-dimensional headers can be made close through proper design. Figure 6.79 shows a schematic of a regenerator-core cross section with inlet and outlet headers. The figure defines relevant nomenclature.

The following are the principal results of the analyses of London, et al. [44]

1. Counterflow is the optimal configuration for oblique-flow headers, because counterflow oblique-flow headers have the smallest pressure drop of any configuration.

2. The velocity profile at the exit of the outlet header is

$$\frac{u_o}{U_o} = \frac{\pi}{2} \sin\left(\frac{\pi}{2}\zeta\right) \; ; \tag{6.214}$$

where

u_o = Local velocity at the outlet (m/s);
U_o = Average velocity at the outlet (m/s); and
$\zeta \equiv y/Y_o$ (-).

3. The inlet and outlet headers must be box-shaped for uniform flow distribution.

4. The height of the inlet header (Y_i) is given by

$$\frac{Y_i}{Y_o} = \sqrt{\left(\frac{4}{\pi^2}\right)\frac{\rho_{sto}}{\rho_{sti}}} \; ; \tag{6.215}$$

where

ρ_{sto}	=	Static density at the outlet (kg/m^3); and
ρ_{sti}	=	Static density at the inlet (kg/m^3).

London, et al. [44] give two criteria for validity of their analyses:

Criteria 1;

$$\frac{\Delta P_{st,m+r}}{\frac{1}{2}\rho_{sti}U_i^2} > \frac{1}{2} \; ; \tag{6.216}$$

where

$\Delta P_{st,m+r}$	=	Pressure drop across the core and any partial screen (Pa);
ρ_{sti}	=	Static density of the gas at the inlet (kg/m^3); and
U_i	=	Average velocity at the inlet (m/s).

Criteria 2;

$$\frac{Y_i}{L} < \frac{1}{3} \; . \tag{6.217}$$

We use the above results and criteria to calculate the core and header pressure drops required for uniform flow distribution.

There is a conflict between designing a heat exchanger for good heat transfer and small pressure drops for cases in which the two flows are at substantially different pressures, because on the high-pressure side the core pressure drop may be too low to produce uniform flow distribution. Wilson [78] has proposed the use of a distributing partial screen to produce good flow distribution in low-pressure-drop cases.

The static-pressure drop across the core is

$$P_{sti} - P_{sto} = \Delta P_{st,m+r} \; ; \tag{6.218}$$

where $\Delta P_{st,m+r}$ is the sum of the static-pressure drops across the core and across a partial fluid-flow screen that would be used to satisfy the criterion given by Eq. 6.216. Section 5.1 describes how to calculate the pressure drop across a regenerator core.

The normalized squared-mean velocity at the exit of the outlet header can be calculated using Eq. 6.214:

$$\frac{\overline{u_o^2}}{U_o^2} = \frac{\pi^2}{4}\int_0^1 \left(\sin\frac{\pi}{2}\zeta\right)^2 d\zeta \; , \tag{6.219}$$

or

$$\frac{\overline{u_o^2}}{U_o^2} = \frac{\pi^2}{8} \; . \tag{6.220}$$

Conservation of mass and Eq. 6.215 are used to obtain

$$\left(\frac{U_o}{U_i}\right)^2 = \frac{4}{\pi^2}\left(\frac{\rho_{sti}}{\rho_{sto}}\right) \; . \tag{6.221}$$

Eqs. 6.220 and 6.221 are used to calculate the squared-mean velocity at the outlet:

$$\overline{u_o^2} = \left(\frac{U_i^2}{2}\right)\left(\frac{\rho_{sti}}{\rho_{sto}}\right) \; . \tag{6.222}$$

The pressure drops across the headers, the core, and the partial screen result in a total-pressure drop from the inlet to the outlet. The following is an algorithm for calculating this total-pressure drop.

1. Calculate the static temperature at the inlet (T_{sti}) from the total temperature (T_{0i}), the average inlet velocity (U_i), and the specific heat capacity at constant pressure ($\overline{c_{Pi}}$): $T_{sti} = T_{0i} - U_i^2/(2\overline{c_{Pi}})$ where $\overline{c_{Pi}} \equiv c_P[(T_{sti} + T_{0i})/2]$;

2. Calculate the static pressure (P_{sti}) at the inlet from the total pressure: $P_{sti} = P_{0i}\,(T_{sti}/T_{0i})^{\overline{c_{Pi}}/R}$;

3. Calculate the gas density in the inlet header: $\rho_{sti} = P_{sti}/(RT_{sti})$;

4. Calculate the average total temperature from the total temperatures at the inlet and outlet: $\overline{T_0} \equiv (T_{0i} + T_{0o})/2$;

5. Calculate an average viscosity for the gas: $\mu_m = \mu_{\text{air}}(\overline{T_0})$;

6. Calculate an average specific heat capacity for the gas: $c_{Pm} = c_{P_{\text{air}}}(\overline{T_0})$;

7. Guess at the total pressure at the outlet, P_{0o};

8. Calculate an average total pressure: $\overline{P_0} \equiv (P_{0i} + P_{0o})/2$;

9. Calculate an average density: $\rho_m = \overline{P_0}/(R\overline{T_0})$;

10. Calculate the pressure drop across the core (ΔP_{stm}) using the method given in Section 5.1;

11. Employ a partial screen if the core pressure drop is too small, and calculate the combined pressure drop of the partial screen and the core accordingly: $\Delta P_{st,m+r} = \max\left[(0.3)\rho_{sti}U_i^2, \Delta P_{stm}\right]$;

12. Guess at the average squared velocity at the outlet, $\overline{u_o^2}$;

13. Calculate the static temperature at the outlet: $T_{sto} = T_{0o} - \overline{u_o^2}/(2\overline{c_{Po}})$ where $\overline{c_{Po}} \equiv c_P[(T_{sto} + T_{0o})/2]$;

14. Calculate the static pressure at the outlet: $P_{sto} = P_{0o}\left(T_{sto}/T_{0o}\right)^{\overline{cp_o}/R}$;
15. Calculate the gas density at the outlet: $\rho_{sto} = P_{sto}/(RT_{sto})$;
16. Calculate the average squared velocity at the inlet: $\overline{u_o^2} = (U_i^2/2)(\rho_{sti}/\rho_{sto})$;
17. Go to step 12 until convergence;
18. Calculate the static pressure at the outlet: $P_{sto} = P_{sti} - \Delta P_{st,m+r}$;
19. Calculate the total pressure at the outlet: $P_{0o} = P_{sto}\left(T_{0o}/T_{sto}\right)^{\overline{cp_o}/R}$; and
20. Go to step 7 until convergence.

Appendix A

Performance of the Ericsson Cycle

Chapter 3 shows that intercoolers and reheat combustors can improve the performance of gas turbines. Cooling the air during the compression process with intercoolers reduces the power requirement of the compressor stages. Heating the gas during the expansion process with reheat combustors increases the power output of the turbine stages. Thus, intercoolers and reheat combustors can improve thermal efficiency and cycle specific power. A gas-turbine cycle with an infinite number of intercoolers and an infinite number of reheat combustors is known as an Ericsson cycle, after the Swedish inventor John Ericsson. This appendix shows that the thermal efficiency of an Ericsson cycle with ideal components approaches the Carnot efficiency, which is the maximum thermal efficiency achievable by any heat engine.

To find the thermal efficiency of an Ericsson cycle, we will derive expressions for the heat inputs of the combustors, the power inputs to the compressors, and the power outputs of the turbines. Then, we will let the number of compressor stages, turbine stages, and combustors approach infinity and show that the thermal efficiency approaches the Carnot efficiency.

Thermal efficiency can be calculated as the ratio of the cycle specific power to the cycle specific heat-input rate:

$$\eta_{TH} = \frac{W'}{Q'} , \tag{A.1}$$

where

η_{TH}	=	Thermal efficiency (–);
Q'	=	Cycle specific heat-input rate (–); and
W'	=	Cycle specific power (–).

Cycle specific power can be calculated using Equations 3.5 and 3.6. Cycle specific heat-input rate is the sum of the specific heat-input rates of the n

combustors in the cycle:

$$Q' = \sum_{j}^{n} Q'_{Hj} \,, \tag{A.2}$$

The specific heat-input rate of each combustor is the heat-input rate to the combustor divided by (roughly[1]) the enthalpy flow rate into the gas turbine:

$$Q'_{Hj} = \frac{\dot{Q}_{Hj}}{(\dot{m}c_P T)_0} \,, \tag{A.3}$$

where

Q'_{Hj}	=	Specific heat-input rate of Combustor j (–);
$\dot{Q}_{Hj}$	=	Heat-input rate to Combustor j (W);
$(\dot{m}c_P T)_0$	=	Roughly the enthalpy flow rate into the gas turbine (W);
$\dot{m}$	=	Mass flow rate (kg/s);
c_P	=	Specific heat capacity evaluated at a constant pressure (J/kg-K); and
T	=	Total temperature (K).

With an ideal regenerator (100% effective), the heat-input rate to each of the combustors (including the first combustor) is equal to the power output of each successive turbine stage. Thus, thermal efficiency is

$$\eta_{TH} = 1 + \frac{\sum_{j=1}^{n} W'_{Cj}}{\sum_{j=1}^{n} W'_{Ej}} \,. \tag{A.4}$$

where

$\sum_{j=1}^{n} W'_{Cj}$	=	Sum of the specific powers of the n compressor stages (–); and
$\sum_{j=1}^{n} W'_{Ej}$	=	Sum of the specific powers of the n turbine stages (–).

In this approximate analysis, we assume a constant specific heat capacity of the working fluid (air and combustion products for open-cycle gas turbines). The specific heat capacity of air actually increases by about 10% for every 600 K of temperature increase. For a constant specific heat capacity and ideal compressor stages (100% efficient), the sum of the specific powers of the n compression stages is

$$\sum_{j=1}^{n} W'_{Cj} = n\left[1 - r^{\frac{(R/c_P)}{n}}\right] \,, \tag{A.5}$$

where

r	=	Cycle pressure ratio, the ratio of the outlet pressure of the last compressor stage to the inlet pressure to the gas turbine (–); and
R	=	Gas constant for air (J/kg-K).

[1] As explained in Chapter 3, the denominator in Equation A.2 is only roughly the enthalpy flow rate into the gas turbine because specific heat capacity varies with temperature.

Similarly, the sum of the specific powers of the n ideal (100% efficient) turbine stages is

$$\sum_{j=1}^{n} W'_{Ej} = nT'\left[1 - r^{\frac{-(R/c_P)}{n}}\right] , \tag{A.6}$$

where

T' = Cycle temperature ratio, the ratio of the outlet temperature of each of the n combustors to the inlet temperature to the gas turbine (–).

We insert the expressions for the specific powers of the compressor and turbine stages (Equations A.5 and A.6) into our expression for the thermal efficiency (Equation A.4) and take the limit as the number of compressor stages, turbine stages, and combustors approaches infinity:

$$\eta_{TH} = \lim_{n\to\infty} 1 + \frac{\left[1 - r^{\frac{(R/c_P)}{n}}\right]}{T'\left[1 - r^{\frac{-(R/c_P)}{n}}\right]} = 1 - \frac{1}{T'} \,. \tag{A.7}$$

This is the Carnot efficiency (see Equation 3.3).

References

[1] Babcock & Wilcox Co., "Steam, Its Generation and Use," New York, 1963.

[2] Bahnke, G. D. and C. P. Howard, "The Effect of Longitudinal Heat Conduction on Periodic-Flow Heat Exchanger Performance," Transactions of the ASME, Journal of Engineering for Power, Vol. 86, 1964.

[3] Bannister, R. L., N. S. Cheruvu, D. A. Little, and G. McQuiggan, "Development Requirements for an Advanced Gas Turbine System," ASME Paper 94-GT-388, ASME New York, New York, 1994.

[4] Beck, D. S. "Optimization of Regenerated Gas Turbines," ASME Paper 94-GT-231, accepted for publication in the Transactions of the ASME, Journal of Engineering for Gas Turbines and Power.

[5] Beck, D. S. "Regenerator Effectiveness during Transient Operation," ASME Paper 94-GT-230, accepted for publication in the Transactions of the ASME, Journal of Engineering for Gas Turbines and Power.

[6] Beck, D. S. "Regenerators for Gas-Turbine Applications," PhD thesis in the department of mechanical engineering at MIT, 1993.

[7] Beck, D. S. "The Effect of Seal Width on Regenerator Effectiveness," Transactions of the ASME, Journal of Engineering for Gas Turbines and Power, Vol. 116, No. 3, July 1994.

[8] Beck, D. S. "The Compactness of Rectangular-Passage Regenerators," internal memorandum, department of mechanical engineering, MIT, Cambridge, Massachusetts, 1992.

[9] Bowden, A. T. and W. Hryniszak, "The Rotary Regenerative Air Preheater for Gas Turbines," Transactions of the ASME, July 1953.

[10] Cadwell, R. G., et al., "The Ford Turbine—An Engine Designed to Compete with the Diesel," SAE Paper No. 720168, Detroit, Michigan, 1972.

[11] Carnot, Sadi, *On the Motive Power of Heat.*

[12] Chappell, M. S. and E. P. Cockshutt, "Gas Turbine Cycle Calculations: Thermodynamic Data Tables for Air and Combustion Products for Three Systems of Units," NRC, Aeronautical Report LR-579, National Research Council of Canada, August, 1974.

[13] Chiprich, J. M., "ICR Program Overview," The Institute of Marine Engineers, Eastern USA Branch, 15 February 1994.

[14] Coppage, J. E. and A. L. London, "The Periodic-Flow Regenerator—A Summary of Design Theory," The Transactions of the ASME, Vol. 75, 1953.

[15] "Cowper," an entry in The New Encyclopaedia Britannica, 15th Edition, University of Chicago, Encyclopaedia Britannica, Inc., 1980.

[16] Cox, M. and R. K. P. Stevens, "The Regenerative Heat Exchanger for Gas-Turbine Power Plants," Proceedings of The Institute of Mechanical Engineers, Vol. 163, W. E. P. No. 60, London, UK, 1950.

[17] Davis, James E. and William N. Gill, "The Effects of Axial Conduction in the Wall on Heat Transfer with Laminar Flow," International Journal of Heat and Mass Transfer, Vol. 13, Pergamon Press, 1970.

[18] Day, Paul J. "ATTAP Extruded Ceramic Regenerator Disk Development," Proceedings of the Annual Automotive Technology Development Contractors' Coordination Meeting, 1991 P-256.

[19] Day, Paul J., "A Rotary Heat Exchanger for Automotive and Other Ground-Based Gas Turbines," ASME Paper 94-GT-124, ASME New York, New York, 1994.

[20] Edelman, H. C. M., "In Search of Stirling," *The Announcer*, pp. 6–12, N. V. Philips, Eindhoven, 1969.

[21] Gale, W. K. V., "Ferrous Metals," in *An Encyclopaedia of the History of Technology*, McNeil, Ian (ed.), Routledge, London and NY, 1990.

[22] Grossman, D. G., and J. G. Lanning, "Aluminous Keatite—An Improved Rotary-Ceramic-Regenerator Core Material," ASME Paper 77-GT-60, ASME New York, New York, 1977.

[23] Hagler, Carla D., "The Design of a Ceramic Rotary Regenerator for a Regenerated Low-Pressure-Ratio Gas-Turbine Engine," M. S. thesis, Mechanical Engineering, MIT, 1987.

[24] Hahnemann, H. W., "Approximate Calculation of Thermal Ratios in Heat Exchangers Including Heat Conduction in Direction of Flow," National Gas Turbine Establishment Memorandum No. M36, 1948.

[25] Hansson, Sven A., "Birger and Fredrik Ljungstrom—Inventors," AB International STAL Company, Stockholm, Sweden, 1955.

[26] Harper, D. B., "Seal Leakage in the Rotary Regenerator and Its Effect on Rotary-Regenerator Design for Gas Turbines," The Transactions of the ASME, February, 1957.

[27] Harper, D. B. and W. M. Rohsenow, "Effect of Rotary Regenerator Performance on Gas-Turbine-Plant Performance," The Transactions of the ASME, July, 1953.

[28] Hawkins, W. J., D. Mathieson, C. J. Bruce, and P. Socoloski, "System Development Test Program for the WR-21 Intercooled Recuperative (ICR) Gas Turbine Engine System," ASME Paper 94-GT-186, ASME New York, New York, 1994.

[29] Helms, H. E., et al., "Ceramic Applications in Turbine Engines," DoE/NASA/0017-6, NASA CR-174715, EOR-11442, Cleveland, OH, 1984.

[30] Helms, H. E., et al., "AGT 100 Advanced-Gas-Turbine Technology-Development Project," Proceedings of the 23rd Automotive Technology Development Contractor's Coordination Meeting, SAE Publication, Warrendale, Pennsylvania, 1986.

[31] Huebner, G. J., Jr., personal communication with one of the authors: letter dated 11 January 1994 to D. G. Wilson from Ann Arbor.

[32] Huebner, G. J., Jr., "Chrysler's Automotive Gas Turbine," (approximate title), seminar and automobile demonstration given at MIT, Cambridge, Massachusetts, 1968.

[33] Hirschkron, R. and C. J. Russo, "Small Turboshaft/Turboprop-Engine Technology Study," AIAA/ASME/SAE/ASEE 22nd Joint Propulsion Conference, Huntsville, Alabama, 1986.

[34] Johnson, Dag, and R. J. Mowill, "Aegidius Elling—A Norwegian Gas-Turbine Pioneer," Norwegian Technical Museum, Oslo ,1968.

[35] Kawasaki, K. T. Matsuhisa, I. Sakai, and K. Hijikata, "Heat Transfer Characteristics of Rotating Ceramic Regenerators—Numerical Solution Using a Hybrid Finite Difference/Laplace Transform Scheme," ASME report 91-GT-371.

[36] Kays, W. M. and M. E. Crawford, *Convective Heat and Mass Transfer, Second Edition*, McGraw-Hill, New York, 1980.

[37] Kays, W. M. and A. L. London, *Compact Heat Exchangers, Third Edition*, McGraw-Hill Book Company, New York, 1984.

[38] Kays, W. M., A. L. London, and D. W. Johnston, "Gas Turbine Plant Heat Exchangers—Basic Heat Transfer and Flow Friction Design Data," *ASME Research Publication*, 1951.

[39] Kleiner, R. N., L. R. Strasbaugh, and L. E. Coubrough, "Highly Compact Ceramic Recuperator for Engine Applications," ASME Paper 84-GT-50.

[40] Lambertson, T. J.,"Performance Factors of a Periodic-Flow Heat Exchanger," Transactions of the ASME, Vol. 80, 1958.

[41] Lankford, William J., Jr., et al. (ed.) *The Making, Shaping, and Treating of Steel, 10th Edition*, US Steel, Pittsburgh, Pennsylvania, 1985.

[42] Lanning, J. G., and D. J. S. Wardale, "The Development of a Glass-Ceramic Axial-Flow Rotary Regenerator," ASME Paper No. 66-GT-107, presented in Zurich, Switzerland, ASME New York, New York, 1966.

[43] London, A. L. "Laminar Flow Gas Turbine Regenerators—The Influence of Manufacturing Tolerances," Transactions of the ASME, Journal of Engineering for Power, Vol. 92, 1970.

[44] London A. L., G. Klopfer, and S. Wolf, "Oblique Flow Headers for Heat Exchangers," Transactions of the ASME, Journal of Engineering for Power, Vol. 90, 1968.

[45] London, A. L., W. E. Mason, and L. M. K. Boelter, "Performance Characteristics of a Mechanically Induced Draft, Counterflow, Packed Cooling Tower," Transactions of the ASME, 1940.

[46] London, A. L., D. F. Sampsell, and J. G. McGowan, "The Transient Response of Gas Turbine Plant Heat Exchangers—Additional Solutions for Regenerators of the Periodic-Flow and Direct-Transfer Types," Transactions of the ASME, Journal of Engineering for Power, Vol. 86, 1964.

[47] London, A. L. and R. E. Seban, "A Generalization of the Methods of Heat Exchanger Analysis," Pioneer Paper, International Journal of Heat and Mass Transfer, Vol. 23, pp. 5-16, Pergamon Press Ltd. 1980.

[48] Luikov, A. V., V. A. Aleksashenko, and A. A. Aleksashenko, "Analytical Methods of Solution of Conjugated Problems in Convective Heat Transfer," International Journal of Heat and Mass Transfer, Volume 14, Pergamon Press, 1971.

[49] Matsuhisa, K. K. T. I. Sakai, and K. Hijikata, "Heat Transfer Characteristics of Rotating Ceramic Regenerators—Numerical Solution Using a Hybrid Finite Difference/Laplace Transform Scheme," ASME Paper 91-GT-371.

[50] Mills, A. F., *Heat Transfer*, Irwin, Boston, MA, 1992.

[51] Mondt, J. R., "Correlating the Effects of Longitudinal Heat Conduction on Exchanger Performance," a chapter in *Compact Heat Exchangers—History, Technological Advancement and Mechanical Design Problems*, edited by R. K. Shah, C. F. McDonald, and C. P. Howard, The ASME, New York, New York, 1980.

[52] Mordell, D. L., "An Experimental Coal-Burning Gas Turbine," Proceedings of The Institute of Mechanical Engineers, Vol. 169, No. 7, London, UK, 1955.

[53] Motor, "Power Without Pistons—The Rover B. R. M., Description and Road Test," UK, 25 Sept. 1965.

[54] Mullins, P., "New Naval Propulsion System on Target for 1994 Debut," Diesel and Gas Turbine Worldwide, March 1994.

[55] Penny, Noel, "The Development of the Glass Ceramic Regenerator for the Rover 2S/150R Engine," SAE Paper 660361.

[56] Penny, Noel, personal communication with one of the authors: letter dated 22 December 1993 to D. G. Wilson, Coventry, UK, 1993.

[57] Potter, M. C. and J. F. Foss, *Fluid Mechanics*, Great Lakes Press, Inc., Okemos, Michigan, 1982.

[58] Ogata, Katsuhiko, *Modern Control Engineering*, Prentice-Hall, Inc., Englewood Cliffs, N. J., 1970.

[59] Reese, Jacob, "Regenerating Metallurgic Furnaces," Transactions of the ASME, Vol. 1, 1880.

[60] Sayama, J. and T. Morishita, "Development of a Regenerator for an Automotive Gas Turbine Engine," ASME Paper 92-GT-27.

[61] Shah, R. K. "A Correlation for Longitudinal Heat Conduction Effects in Periodic-Flow Heat Exchangers," a technical brief in The Transactions of the ASME, Journal of Engineering for Power, Vol. 97, Series A, 1975.

[62] Shah, R. K. "Thermal Design Theory for Regenerators," in *Heat Exchangers: Thermal-Hydraulic Fundamentals and Design* edited by S. Kakaç, A. E. Bergles, and F. Mayinger, Hemisphere Publishing Corporation, 1981.

[63] Shah, R. K. and A. L. London, "Laminar Flow Forced Convection in Ducts," *Advances in Heat Transfer*, Academic, New York, 1978.

[64] Shah, R. K. A. D. Kraus, and D. Metzger, *Compact Heat Exchangers, a Festschrift for A. L. London*, Hemisphere Publishing Corporation, New York, 1990.

[65] Sheylag, V. R., "Calculation of the Temperature Field in Plane Slotted Channels with Laminar Gas Flow," Journal of Engineering Physics, Volume 12, 1967.

[66] Shepard, S. B., T. L. Bowen, and J. M. Chiprich, "Design and Development of the WR-21 Intercooled Recuperated (ICR) Marine Gas Turbine," ASME Paper 94-GT-79, ASME New York, New York, 1994.

[67] Shimada, K., et al., "Advanced Ceramic Technology Developed for Industrial 300 kW Ceramic-Gas-Turbine Research and Development Project in Japan," ASME Paper 93-GT-188, ASME New York, New York, 1993.

[68] "Siemens," an entry in The New Encyclopaedia Britannica, 15th Edition, University of Chicago, Encyclopaedia Britannica, Inc., 1980.

[69] Singh, S. N. "Heat Transfer by Laminar Flow in a Cylindrical Tube," in *Applied Science Research*, series A, Vol. 7, 1957.

[70] Wadman, B., "Diesel and Gas Turbine Programs at Westinghouse Marine," Diesel and Gas Turbine Worldwide, September 1995.

[71] Walker, G., *Stirling Engines*, Clarendon Press, Oxford, 1980.

[72] Wilson, D. G., "A Method of Design for Heat-Exchanger Inlet Headers," ASME Paper 66-WA/HT-41.

[73] Wilson, D. G., *Design of High-Efficiency Turbomachinery and Gas Turbines*, The MIT Press, Cambridge, Massachusetts, 1984.

[74] Wilson, D. G., "The Continuous-Strip Regenerative Heat-Exchanger," preprint 20, AICLE annual meeting, 1961.

[75] Wilson, D. G., "Heat Exchanger Containing a Component Capable of Discontinuous Movement," US Patent No. 5,259,444, 9 November 1993, assigned to MIT.

[76] Wilson, D. G., private communication of draft paper, 1993.

[77] Wilson, D. G., private communication of pending patent and draft paper, 1993.

[78] Wilson, D. G., private communication, 1991.

[79] Wilson, D. G., R. G. Frenkel, D. J. Kowalick, H. M. Nahatis, S. M. Silverstein, and L. A. Tampe, "Coal-Burning Exhaust-Heated-Cycle Gas Turbine with a Regenerative Heat Exchanger," ASME Paper 91-GT-342, ASME New York, New York, 1991.

[80] Wilson, D. G., "Low-Leakage and High-Flow Regenerators for Gas-Turbine Engines," Proceedings of the Institute of Mechanical Engineers, Vol. 207, Part A: Journal of Power and Energy, pp. 195-202. London, UK, 1993.

[81] Wilson, D. G., and D. S. Beck, "Preliminary-Design Considerations for a Hybrid-Vehicle Small Gas Turbine," a progress report, Department of mechanical engineering, MIT, Cambridge, Massachusetts, January 1993.

[82] Wilson, D. G., "The Supplementary-Fired Exhaust-Heated Cycle for Coal, Wood, and RDF," I. Mech. E. Jrl. of Power and Energy, Vol. 207, pp. 203–8, 1993.

[83] Wilson, D. G., "Automotive Gas Turbines: The Pioneers," IGTI Global Gas Turbine News, May/June 1995.

[84] Wilson, D. G., "Automotive Gas Turbines: Progress and Setbacks," IGTI Global Gas Turbine News, August/September 1995.

[85] Wilson, R. A. and D. B. Kupratis, "Future Vehicular Recuperator Technology Projections," ASME Paper 94-GT-395, ASME New York, New York, 1994.

[86] Zarinchang, J. *The Stirling Engine*, Intermediate Technology Publications, London, 1972.

Index

Aircraft gas turbines, 46, 64
Allison, 64
Annular-sector shaped seals, 20
Automobiles, 39
Automotive gas turbines, 46, 50, 80, 120
Axial-flow regenerator, 2
Axial-flow regenerators, 68, 80, 98, 99, 140

Bahnke and Howard, 13
Bahnke, G. D. , 34, 92, 114, 137, 140, 143, 189
Baths, 27
Boelter, L. M. K. , 33
Boiler efficiency, 30

Carnot efficiency, 38, 46, 62, 235, 237
Carnot, Sadi, 38
Carry-over leakage, 22, 64, 84, 87, 122, 137, 227
Ceramic Applications in Turbine Engines (CATE), 64
Ceramic-disk cores, 80
Ceramics, 17, 32, 47, 50, 67, 88, 120, 167
Chappell, M. S. , 41, 49, 91
Checkerwork, 28
Chrysler, 30
Clamping seals, 20
Clothing, 27
Coal burning, 76
Cockshutt, E. P. , 41, 49, 91
Cold start-ups, 17, 168
Combined-cycle engines, ix
Combustion, 48
Combustion products, 92
Combustor, ix
Combustors, 235
Compact heat exchangers, 25, 34, 98, 122, 140
Compactness, 2, 9
Compression, 45, 55
Compressors, ix, 235
Conduction, 10, 34, 86, 87, 122, 124, 140, 141
Conduction under seals, 13
Conservation of energy, 126
Conservation of mass, 232
Conservation of momentum, 131
Constant, Hayne, 30
Convection, 124
Convective conductances, 7, 84
Convective heat-transfer coefficient, 15
Convective-conductance ratio, 8, 94
Cooling towers, 33
Coppage, J. E. , 34, 93
Cordierite, 88
Core Compactness, 9, 25, 95
Core rotation rate, 102, 133
Core rotation speed, 99
Core specimens, 174
Core volume ratio, 80
Core-conduction parameter, 12
Core-Rotation Effect, 85, 139
Cores, ix
Corning, 32, 67
Correlations, 35, 93
Correlations of effectiveness, 121, 123
Cost, 2, 50
Cost of a regenerator, 101
Counterflow headers, 230

Counterflow recuperator, 121, 137
Cowper stove, 28, 69
Cowper, Edward, 28
Crawford, M. E. , 88, 117, 125
Cycle calculations, 37, 40, 101
Cycles, 37

Department of Energy (DoE), 64
Development length, hydraulic, 124, 125
Development length, thermal, 124, 125
Diameter of a core, 87, 102
Diesel engines, ix, 46, 50
Dimensionless core rotation rate, 84
Dimensions, 96
Direct leakage, 64, 227
Direct Regenerator Design, x, 79
Direct seal leakage, 22
Discontinuous rotation, 20, 69, 201
Distribution of flow, 84, 99, 229
Dynamics, 17, 161

Effect, 142
Effective convective heat-transfer coefficient, 16
Effectiveness, 4, 80, 102, 121, 170
Effects, 122
Efficiency, ix, 41, 49
Elling, Aegidius, 30
Energy equation, 154
Energy exchanger, 33
Ericsson cycle, 62, 235
Ericsson, J. , 235
Exhaust-gas heat exchange, ix, 1, 30
Exhaust-gas heat exchanger, 63
Exhaust-heated cycle, 76
Expansion, 49, 56
Experience in regenerator design, 101
Experimentation, 173
External combustion, 48
Extrusion, 19

Finite-difference numerical integration, 137, 140, 153, 199
Finite-difference numerical-integration methods, 34
First law of thermodynamics, 40
Flow areas, 84, 96
Flow exposure, 10, 15, 18
Flow-friction data, 87
Ford, 33, 64
Ford 704 gas turbine, 56
Foss, J. F. , 125
Fourier number, 14, 142, 156
Fourier's law, 11
Free convection, 124
Friction coefficient, 87, 98
Fully developed flow, 125

Garrett, 64
Gibbs' equation, 41
Governing equations, 126
Ground-based gas turbines, 64

Hagler, C. D. , 68
Harper, D. B. , 34, 86, 99, 101, 227
Headers, 18, 25, 35, 50, 229
Heat balance, 12
Heat capacity, 87
Heat diffusion, 13, 15, 141
Heat transfer, x, 4, 121
Heat-capacity rates, 5, 93
Heat-capacity-rate ratio, 6, 94, 137
Heat-Recovery Boiler (HRB), 50, 59
Heat-Recovery Steam Generators (HRSG's) 50
Heat-transfer data, 87
Heat-transfer parameters, 92, 94
Height of a Transfer Unit, 33
Helicopter gas turbines, 64
Helms, H. E. , 64
High-pressure compressor, 1
Hirschkron, R. , 64, 80
Howard, C. P. , 34, 92, 114, 137, 140, 143, 189
Hryniszak, Waldo, 30
Huebner, George J., 30
Humps, 223
Hydraulic diameter, 88, 98, 125

Intercooled Regenerative (ICR) cycle, 55

Intercooled Regenerative (ICR) gas turbine with reheat, 102
Intercooler, ix
Intercooling, 1, 55, 235
Internal combustion, 48
Isentropic, 46

Kays and London, x
Kays, W. M. , 2, 12, 34, 88, 93, 105, 112, 114, 117, 125, 137, 139, 140, 172, 198

Lambertson, T. J. , 34, 92, 137, 188, 195, 198
Laminar flow, 9, 25, 87, 95, 131
Le Mans, 32
Leakage, x, 2, 22, 30, 33, 48, 64, 80, 86, 87, 99, 102, 122, 227
Ljungstrom rotary air preheater, 30, 67
Ljungstrom, Fredrik, 30
London, A. L. , 2, 12, 33, 35, 88, 93, 105, 112, 114, 117, 125, 137, 139, 140, 172, 198, 229
Losses, 39, 43, 62
Low cost, ix

Maintenance, ix
Maldistribution of a flow, 229
Manufacturing, 2, 10, 15, 19, 32, 35, 67
Manufacturing processes, 28
Mason, W. E. , 33
Matlab, 156
Mesh Screen Matrix (MSM) regenerators, 80, 86
Metallic blade, 46
Metallurgic furnaces, 30
Minimum core volume, 84
Modular regenerator, 71, 149
Momentum equation, 153
Mondt, J. R. , 13
Mordell, D. , 76

NASA, 64
National Gas-Turbine Establishment (NGTE), 30

Negative core rotation, 227
Newton's laws, 5, 10, 47, 126
Non-uniform passages, 19
Non-uniformity flow distribution, 18
Non-uniformity of flow, 180
Number of Transfer Units (NTU), 8, 95
Numerical procedures, 92, 121
Numerical-graphical methods, 34
Nusselt number, 10, 87, 98, 155

Oblique-flow headers, 229
Office of Naval Research, 34
One-dimensional heat diffusion, 141, 144
Open-hearth furnace, 28
Optimal Design, ix, 34
Optimal Regenerator Design, x, 101
Optimization parameters, 102
Outputs of Direct Regenerator Design, 99

Péclét number, 125
Parallel-plate passage geometry, 15
Parallel-plate passages, 80, 131
Partial screen, 84, 99, 231
Passage geometries, 9
Passage tube, 7, 15
Passage tubes, 124
Pebbles, 27
Penny, Noel, 32
Performance, x
Performance of a core, 95
Permeability, 9, 25, 95
Pollution, ix, 50
Porosity, 23, 87, 151
Porosity number, 16, 151
Positive core rotation, 227
Potter, M. C. , 125
Power consumption, x, 4, 26, 80
Power output, ix
Power-generation gas turbines, 46, 80
Prandlt number, 125
Preheat, ix, 1
Preliminary design, ix

Pressure drops, x, 24, 43, 50, 80, 84, 95, 229
Pressure ratio, 64, 102
Pressures, 99
Principle of operation of gas turbines, 43
Process furnaces, 27
Properties, 124, 125
Properties of gases, 49, 91

Radial-flow regenerator, 69
Radial-flow regenerators, 80, 98, 99, 140
Radiation, 124
Rectangular passages, 88
Recuperators, 1, 2, 10, 66, 121, 137
Reese, Jacob, 28
Regenerative cycle, 52
Regenerator, definition, 27
Reheat, 235
Reheat Combustor, 56
Reynolds number, 87, 125
RGT-OPT, 37, 41, 79, 102, 120
Ritz, 30
Rohsenow, W. M. , 34
Romans, 27
Rotary air preheater, 30
Rotary regenerators, 2, 10, 64
Rotation, 10, 34
Rotation of a core, 122, 137, 201, 229
Rotation-period-to-Pause-period Ratio (*RPR*), 201
Rotorcraft, 39
Rotorcraft gas turbines, 80
Rover Cars, 32
Russo, C. J. , 64, 80

Scarf, 27
Scrubbers, 50
Seal clearance, 22, 86
Seal coverage, 14, 86, 102, 122, 140
Seal leakage, 227
Seal length, 102
Seal shape, 19, 180
Seal width, 140
Seal-leakage parameter, 24
Seals, 69, 124
Seban, R. E. , 33
Second law of thermodynamics, 41
Shah, R. K. , 35, 93, 125, 140
Siemens, Friedrich, 28
Siemens, Karl Wilhelm or Sir Charles William Siemens, 28
Simple cycle, 42, 63
Simplifying assumptions, 124
Size of a regenerator, 101, 102
Small-Engine-Component Technology (SECT), 64
Smart seals, 30
Software, 37
Solar-Heated Cycle, 76
Solid areas, 95
Solid-area ratio, 12
Spark-ignition engines, ix, 46, 50
Specific enthalpy, 91
Specific heat capacity, 46
Specific heat-input rate, 235
Specific power, 38, 235
Speed of core rotation, 101
Stainless steel, 167
Stanford University, 34
Static properties, 40
Steady-Flow Energy Equation (SFEE), 40
Steady-state, 17, 40
Steady-state operation, 137
Steam generators, 30
Steam injection, 30
Stirling, Robert, 28
Stones, 27
Sulphur attack, 33
Switching regenerator, 202

Taylor expansion, 154
Temperature profile, 11, 13, 17, 85, 116
Temperature-vs.-entropy plots, 41
Thermal conductivity, 9, 92
Thermal diffusivity, 14
Thermal efficiency, ix, 38, 101, 120, 235

Thermal properties, 88
Thermal stresses, 140
Thickness of a core, 95, 96, 122, 140
Three-dimensional headers, 229
Total properties, 40
Total seal leakage, 22, 227
Transient operation, 17, 168
Transient time period, 17
Truck gas turbines, 56
Turbine blades, 46
Turbine cooling, 46, 63
Turbine Inlet Temperature (TIT), ix, 46, 64
Turbines, 235
Two-dimensional headers, 229

Uniform flow distribution, 35, 122, 229
Uniform flows, 84
Uniform passages, 25, 95
Uniform-width seals, 20

Viscosity, 92

Weight of a regenerator, 101
Wetted perimeter, 10
Width of a core, 87

MIX
Papier aus verantwortungsvollen Quellen
Paper from responsible sources
FSC® C105338

If you have any concerns about our products,
you can contact us on
ProductSafety@springernature.com

In case Publisher is established outside the EU,
the EU authorized representative is:
Springer Nature Customer Service Center GmbH
Europaplatz 3, 69115 Heidelberg, Germany

Printed by Libri Plureos GmbH
in Hamburg, Germany